The Spirit of the Modern System of War

DIETRICH HEINRICH VON BÜLOW
EDITED AND TRANSLATED BY
C. MALORTI DE MARTEMONT

CAMBRIDGE
UNIVERSITY PRESS

CAMBRIDGE
UNIVERSITY PRESS

University Printing House, Cambridge, CB2 8BS, United Kingdom

Published in the United States of America by Cambridge University Press, New York

Cambridge University Press is part of the University of Cambridge.
It furthers the University's mission by disseminating knowledge in the pursuit of
education, learning and research at the highest international levels of excellence.

www.cambridge.org
Information on this title: www.cambridge.org/9781108061575

© in this compilation Cambridge University Press 2013

This edition first published 1806
This digitally printed version 2013

ISBN 978-1-108-06157-5 Paperback

CAMBRIDGE LIBRARY COLLECTION

Books of enduring scholarly value

History

The books reissued in this series include accounts of historical events and movements by eye-witnesses and contemporaries, as well as landmark studies that assembled significant source materials or developed new historiographical methods. The series includes work in social, political and military history on a wide range of periods and regions, giving modern scholars ready access to influential publications of the past.

The Spirit of the Modern System of War

Dietrich Heinrich von Bülow (1757–1807) served for sixteen years in the Prussian army, but for the remainder of his life lived a varied existence as a theatrical manager, preacher, writer, businessman, debtor and finally prisoner. It was not until after his death that this work, written in 1799 and reissued here in its 1806 English translation, began to find favour. He advocates the use of mathematical principles in devising swift and decisive tactics, and was one of the first theoreticians to regard tactics and strategy as separate concerns. At a time when Germany and Italy were yet to be unified, he writes that expansion to 'optimum' size would result in a Europe of only twelve states. The book's theories were criticised by Napoleon and Clausewitz, but had a considerable influence on the Prussian and Austrian armies of the late nineteenth century, and it is regarded as one of the founding texts of modern geopolitics.

Cambridge University Press has long been a pioneer in the reissuing of out-of-print titles from its own backlist, producing digital reprints of books that are still sought after by scholars and students but could not be reprinted economically using traditional technology. The Cambridge Library Collection extends this activity to a wider range of books which are still of importance to researchers and professionals, either for the source material they contain, or as landmarks in the history of their academic discipline.

Drawing from the world-renowned collections in the Cambridge University Library and other partner libraries, and guided by the advice of experts in each subject area, Cambridge University Press is using state-of-the-art scanning machines in its own Printing House to capture the content of each book selected for inclusion. The files are processed to give a consistently clear, crisp image, and the books finished to the high quality standard for which the Press is recognised around the world. The latest print-on-demand technology ensures that the books will remain available indefinitely, and that orders for single or multiple copies can quickly be supplied.

The Cambridge Library Collection brings back to life books of enduring scholarly value (including out-of-copyright works originally issued by other publishers) across a wide range of disciplines in the humanities and social sciences and in science and technology.

THE

SPIRIT

OF THE

MODERN SYSTEM OF WAR.

THE

SPIRIT

OF THE

MODERN SYSTEM OF WAR.

BY

A PRUSSIAN GENERAL OFFICER.

WITH

A Commentary;

BY

C. MALORTI DE MARTEMONT,

KNIGHT OF THE ROYAL AND MILITARY ORDER OF ST. LOUIS, AND
CAPTAIN IN THE FRENCH ROYAL ARTILLERY; MASTER OF
FORTIFICATION AND ARTILLERY IN THE ROYAL
MILITARY ACADEMY AT WOOLWICH.

LONDON:

SOLD BY T. EGERTON, MILITARY LIBRARY, WHITEHALL.

1806.

Printed by C. Mercier and Co.
Northumberland Court, Strand, London.

CONTENTS.

PART THE FIRST.

*A View of the Principle on which the Modern
System of War is founded; of the Alterations
it has produced in the Military System; and of
the Rules resulting from it for the different
Circumstances and Operations of War.*

CONTENTS.

PART THE SECOND.

Consequence of the Principle which requires a Military Base previous to any Operation.

CHAPTER I. Masses, that is to say, the greatest Number of fighting Men, and the greatest Quantity of the Materials requisite in

CONTENTS.

CONTENTS.

PART THE THIRD.

Application of the Principle of the Base to past Military Events, and to those that may take Place in future.

COMMENTATOR'S PREFACE.

The work which I now offer to British Officers in their own language came originally from the pen of a Prussian General, who, if we may depend upon the editor of the bad French translation of it, which, in spite of its innumerable errors, has had an immense circulation in France, unites to the advantage of being distinguished for his military knowledge, that of being honoured with the confidence of a prince of the North of Germany, illustrious for his talents in war, and for his virtues. With respect to myself, I candidly own I am entirely ignorant of the military reputation, connections, and even of the name of the author, who has probably been induced by particular reasons, which in reading his work it is not difficult to guess, to publish his book anonymously*: all that I can say is, that, hearing from different persons of the sensation it had caused on the continent, I procured a copy of it, and thinking it, on perusal, deserving in many respects the praise bestowed upon it, I determined to publish it in English, with a comment of my own.

I am far, however, from always coincid-

---

* I know, from high authorities, that when this work appeared in Prussia, where it excited a great sensation, it was attributed to three different general officers; but which of the three was the author was not positively known. Since then, two French works, in speaking of it, have formally mentioned General Bulow as the author, previously well known by several other highly-esteemed military works.

ing with the author in opinion. The principles which he establishes, as fundamental points of the modern system of war, appear to me almost always consonant to the sound maxims of the art of war, according to the practice of our days; and, in this view, I find the work the more useful, from its treating of those principles in a manner entirely new, and which greatly facilitates the understanding of them; but I cannot see many of his inferences in the same light: for example, to pretend that an art, so subject to political, moral, and physical influences as that of war is and ever will be, whatever system prevails, will one day become, in consequence of the modern system of war, metamorphosed into science, and reduced to invariable rules; to conclude, that all the knowledge relative to this *art*, become *science*, will be circumscribed in sonarrow a sphere, that it will no longer furnish any stimulus to emulation, or excite an ambition to possess it, and that consequently men of genius will prefer directing their attention to objects of more general utility; to affirm that a certain number of Powers, dividing Europe among them, will extend themselves to their natural limits, beyond which they will not dare to expose themselves, and that from this order of things will arise a perpetual peace, the foundation of universal happiness, &c. &c. are some of those dreams created by an ardent imagination, which, hurried away by the impetuosity of its spring, never stops to examine how far calculations, founded on theoretic data, are admissible in practice. Indeed, according to the description which the au-

thor has been pleased to give of a Frenchman*, I should have suspected him to be
my countryman, if I had not been informed that he was a Prussian.

The art of war, certainly, will never
become simply a science. The fundamental principles of it may, indeed, hereafter
be demonstrated in a more simple manner
than has hitherto been done, and by a
method similar or analogous to that of
the author; but, be the existing system
what it may, those principles must ever,
and necessarily, be modified in the application of them, by political, moral, and
physical causes: and men of genius will
always find room, in the profession of
arms, to display and exert the talents they
may have received from nature or acquired
by study, and therefore their ambition to
learn and their emulation will remain the

* See page 276.

same. Allowing, on the other hand, that
Europe should be one day divided among
a certain number of great Powers, and that
by this division those Powers were confined in what the author calls their natural
limits, that event, which by the way I
think very possible, would still not be
followed by perpetual peace, though perhaps it would render wars less frequent;
for, even admitting what the author says,
that after such a division each Power, so
taking its share of Europe, would be circumscribed within the limits prescribed to
it by nature, and beyond which it could
not attempt an offensive war, without
being exposed to all the disadvantages of
an insufficient base, of lines of operation
in danger, &c. &c. the fear of those obstacles would vanish before the influence
which the passions possess over the heart
of man, and the Power that was inclined

to attack another, would flatter itself that it would be able to counterbalance them by manœuvres more ingenious than those of his adversary, or by some other means; now, the passions will exist as long as men exist, and, consequently, while there are men there will be wars.

It is not the less true, however, that the General's work is that of a man of genius; and also, in spite of some systematic notions which appeared to me to admit at least of a discussion, that of a man perfectly acquainted with war, the fundamental principles of which he demonstrates, as I have already said, in a manner entirely new, clear and easily understood: I say fundamental principles, for as to the modifications depending upon local, political, or moral circumstances, genius and the *coup d'œil* are the only masters that can teach them.

I consider this work, then, taken altogether, as being of unquestionable utility; particularly to those officers whose military knowledge is still confined to the circle of Tactics, and who wish to extend it, by making themselves acquainted with the fundamental principles on which (excepting, let me repeat, the modifications depending on local and other circumstances) the offensive and defensive operations of armies are prepared and directed in the modern system of war. The general notions they will here acquire, will unfold their ideas relative to those operations, and even were those notions to be of no other use to them, than that of enabling them to read with advantage the productions of the great masters of the art in the ancient as well as modern system of war *, or such as relate the actions of those masters, that alone would render the work of great moment to them. In fact, throughout all the armies in Europe, let Vegetius, Polybius, Folard, Montecuculli, Turenne, Lloyd, Tempelhoff, the Campaigns of the King of Prussia, and other works more modern, even those which relate to the wars of the French Revolution, be put into the hands of the Officers I am speaking of, and they will most frequently see in them only ideas of Tactics, or historical facts, which will contribute nothing to the enlargement of their

* It may appear extraordinary to some young officers, that in a work intended to explain the modern system of war, I should mention, as an advantage, the being able to read those which relate to the ancient system; but let those who have served long enough to complete their professional knowledge, determine how far a critical examination of the latter contributes to develope the former system.

professional knowledge, because they would not be able to understand the causes which produced them. Such works suppose rather than give the knowledge of the elementary principles, without which it is impossible to analyze them, and sometimes even to understand them. Indeed, those Officers in reading the history of a campaign will see that the armies set out from such and such points; that they marched to such other points; that such movements, such battles, such retreats, &c. took place: but I may still ask what professional advantage will they derive from the knowledge of those facts, if, while they learn them from history, they are not able to reason on the causes which produced them, or to account for them? And could they reason on those causes, if they are unacquainted with the principles on which the operations of armies ought to

be conducted? It is therefore necessary,
before every thing else, to attain a knowledge of those principles, of such at least
as may be considered as fundamental
principles, and from which it is necessary not to deviate, unless induced or
compelled to it by peculiar circumstances, respecting which no rules can
be given, as I have already observed.
Now, I know no works which explain
and unfold those fundamental principles
by so simple and easy a method as that
of the Prussian General.

The author likewise analyzes the most
important operations of Tactics, and dis-
cusses the advantages and defects of those
that are the most generally admitted; he
then takes a view of the principal military
events which have occurred since the introduction of the new system of war
among the European nations; he exa-
mines the most celebrated campaigns,

among others those of the King of Prussia, Frederick II He also reviews those
of Dumourier, in the war of the French
Revolution. His examination furnishes
him with new proofs of the justness of
his arguments; with new grounds for his
assertions; and, from the censure or praise
he bestows on the masters of the art, he
draws instructive lessons, which contribute
to render his work extremely interesting.

Though several of the military combinations of the author have been already
verified, there are still some which coin-
cide less with the events that have taken
place in Europe since the publication of
his work. The late gigantic campaign of
the French seems even to weaken the
force of his reasonings, in the demonstration of the justness of the fundamental
principles of the art of war, as he has
laid them down; but the author, as a
man well informed, has grounded those

reasonings and those principles on probable data; he did not take into his calculations extraordinary incidents, such as
those, for example, to which the campaign
I allude to has given rise, and which,
being too improbable to be taken into any
kind of reasoning, certainly prove nothing,
notwithstanding their results, in opposition to the principles laid down by the
author. I even dare affirm, that the French
would have severely felt the application of
those principles, if a fault in the outset,
occasioned by too much precipitation on
one side, had not rendered the commission
of others inevitable: but it is not my intention to expatiate on this subject.

I have taken upon me to add a commentary; sometimes for the purpose of displaying, in a more circumstantial manner
than the author has done, certain important principles, which, in my opinion, he

has laid down in too general a way; sometimes in order to render his demonstrations clearer and more intelligible, where they appeared to me to require it; and sometimes in order to discuss some of his important positions, of the justness of which I was not sufficiently convinced. From the clashing of opinions light is often struck, and, far indeed from presuming to pronounce my mode of thinking preferable to his, I leave to the reader to decide between us.

One of the assertions of the author, which, on account of the importance of the object to which it relates, and of the objections I think I may justly make to it, seems to me to demand a serious discussion, is, that in order to render the service of the columns of attack ineffectual (*see pages* 174 *and* 175) the adversary must not wait their shock, but withdraw aside of them, so as to menace their flanks. There is no doubt that columns can break only troops before them, and that an enemy, by taking a side position, may avoid their shock; but would he not in so doing expose himself to a greater danger than that of direct resistance? This question I propose to examine here*.

But, before I enter fully on this subject, it is necessary to examine another proposition of the author's. He says, it is no great misfortune for an army to be broken in its centre. I beg leave to refer the reader to page 105 for the demonstration, which would doubtless be a very satisfactory one, if it were to be supposed

* According to the plan I had laid down, and which I have generally followed, the commentaries should all be annexed to the passages to which they relate, in the body of the work; particular circumstances, however, of no importance to the reader, have obliged me to introduce some here. I beg his indulgence for this irregularity, which it has not been in my power to avoid.

that the enemy cd (*fig*. 33, *as referred to
in the text*) had attempted to break the
army A B in the centre, without having
taken precautions to check the wings **A**
and B, and that, after succeeding in his
attempt, he inconsiderately resolved to advance; but such a supposition is not to
be entertained, and it is clear that $c\,d$
would not fail, before the attack, to place
corps opposite to each of those wings, in
order to prevent their movements against
the flanks of the attacking troops: nay, he
might have had it in his power so to place
those corps, as to make the wings fear
being turned by them; and, in either
case, it seems evident to me that, during
the wheeling which the separated parts e, f
of A B would have to make, and which
it would be very long for the troops nearest the centre to execute, even according
to the method of changing front, explain-

ed by the author, page 128*, those separated parts would be very much exposed
to be taken in flank, and perhaps in rear,
by the corps of the wings, which would
be enabled, by a very short movement, to
execute this manœuvre. Now, does not
the author himself, again and again, in the
course of his work, tell us the imminent
danger of an army, when turned on its
flanks, even by a very inferior force? I
know that e and f will not execute their
wheeling without the precaution of having their flanks protected by troops; but
if those troops are defeated, and the vic-

---

* This method seems to me well adapted to a number
of troops not very considerable, and the French in the
last war often successfully made use of the *running step.*
But I much doubt its being equally adapted to a line of
great extent, and it is to be regretted that the author
omitted to mention the numerical force of the raw
troops, whom he thus saw change their front *in the
twinkling of an eye,* which it would be very important
to know.

tors, pursuing rapidly their advantage, come up with e, f, while they are manœuvring, how will they effect their eccentric retreat? Besides, cannot $c\,d$ have them attacked at the same time in front, were it only by *Tirailleurs?* By what means then could e, f, whose front and flanks would be assailed at once, prevent $c\,d$ from advancing? And if he advances, how will they be able afterwards to unite again, as the author says, in a direction parallel to their former front? For, besides that $c\,d$, who moves perpendicularly to his front, has great advantages, by which he can anticipate e, f, at the point where they would attempt their junction, it is also to be remarked, that they cannot effect it without exposing the flank one way or other; for, if they face the wing corps, $c\,d$ will then be upon their flank. It seems to me evident then, that e, f, will be cut off from each other, if $c\,d$

knows how to act; and, if they are cut off, what terrible consequences have they not to fear? May not *c d,* whom every thing favours, occupy awhile in front of *f,* for example, a defensive position, which, by enabling him to advance superior forces against *e,* may furnish him with the means of completely defeating the latter? May he not afterwards execute the same manœuvre with respect to *f;* and if, in consequence of these defeats, *e* and *f* are pushed to a greater distance from each other, which would probably be the case, may not their lines of operation, nay, even their magazines, be highly exposed? I therefore think, that an army, broken in the centre, should not attempt to retreat eccentrically; and more, that an eccentric retreat, in general, is an operation very dangerous to the army that undertakes it; unless, from local and other circumstances, it is completely secured against the movements of the enemy on its flanks. I acknowledge, however, that this kind of retreat does not seem to me to be attended with the same danger in Strategics as in Tactics, because the distance at which the army then is from the enemy, whose movements are observed, renders it easier to manœuvre in time, so as to secure its flanks, and prevent the enemy from advancing; I shall, nevertheless, observe, that if an army retreats eccentrically in many divisions, as shown in *fig.* 28, an active and manœuvring enemy may, without any great difficulty, find means to cut off the divisions *f, k* of the wings, before those that are next to them can come up to their assistance.

With respect to what the author says, that an attack on the centre is contrary to the spirit of the modern system of war, he is certainly right, and if it ever is excusable to attempt so dangerous an enterprise, it is only when the enemy's centre, being much weaker than his wings against which nothing can be undertaken, and it being at the same time impossible to turn his flanks, there is, besides, a certainty, that nothing is to be feared from those wings. I now return to the method proposed by the author for rendering ineffectual the attempts of the columns of attack.

It is to be regretted that the author did not enter into more circumstantial particulars respecting this method: in the first place, because the movement he points out may be conceived in different ways; and, in the second place, because the vague manner in which he states the distance from the enemy, at which it is to be made, leaves, in my opinion, much to be desired. As there appears to me, however, only three ways of executing this movement, I will examine all the three successively.

First, I will suppose that the whole army is to be thrown back into a single line, oblique in respect to its first front; and here I must observe, in the first place, that, owing to the great extent of modern armies in general, a manœuvre of this kind would require the more time, in proportion as the angle formed by the old and new line should be more or less open. The ground, too, might obstruct the marching of the troops, and give rise to difficulties, which, considering the nearness of the enemy, and that the whole army is in motion, would probably be attended by bad consequences. The author, indeed, says, that in flat countries, the movement of the army should be covered by the cavalry; but if this cavalry is defeated by that of the enemy, and flies while the army is still manœuvring, it will not only cease to cover it, but may possibly take it along with it in its flight.

I must also observe, that in throwing itself back into a line, oblique to its first front, the army would most probably expose the flank near the enemy; for, the point of APPUI which covered that flank when the line occupied its former position may not be so advantageous to it in the latter; in which case, as the army would not have sufficient time to procure a new point of APPUI by means of the ground which, besides, may perhaps not afford a sure one, or to supply the want of it by the resources of art, such as WORKS, and ABBATIS, it would have no alternative but to protect its flank by a body of troops: but it is the more to be feared that this step would be insufficient, as it is to be presumed that the enemy, in forming his order of battle, takes care to be able to march troops rapidly to such points, where, according to circumstances, they may be wanted; and that he will thus have it in his power to oppose superior forces to those detached by the army to cover its flank. Now, if that flank be once uncovered and successfully attacked, as probably it would be, would not the army, checked likewise in front by the enemy, find itself in a position quite as critical as that in which it would have been, if broken by the columns. It would derive no advantage then from having avoided their shock.

Secondly, If we now suppose the defensive army to separate into two divisions, one to the right and the other to the left, in order to threaten at the same time the two flanks of the enemy, it would then be subject to the same inconveniences as those which I have deduced from the case of an army being broken in the centre: in either supposition, the flanks of the defensive army will be alike exposed.

Lastly, If, as is seen in *fig.* 59*, only the part *c* B of the army A B, which is opposite to the columns of attack *e*, *f*, is thrown into the oblique line *c d*, bearing its flank to the point *c*, in order to protect at the same time that point, which, on account of the movement of *c* B, becomes the flank of the portion of the line which remains still, this manœuvre will lead to nothing ; for, to render it abortive, it is enough that the columns change their direction, and move against some points near *c*, which will not lengthen their march so as to render it impossible. If even this

---

* Figure 59 does not belong to the author's work, and I only introduced it as it appeared to me necessary to illustrate my own demonstration.

point *c*, which is feebly protected by the
musketry, being only defended by oblique
fires, was not reinforced by the effect of
a powerful artillery, the column *e* would
march rapidly against it, and, having penetrated, would favour the movements of
the troops, which would certainly advance
to support that column, and complete the
affair. I say complete the affair, for, when
once the assailants have made their way
through, the lines A *c* and *c d* would both
of them be taken in flank and rear.

It is only in the most general points of
view, indeed, that I have examined the author's opinion relative the mode of securing an army against the effect of columns
of attack*, and no doubt there are inci-

---

* The author will likewise say, page 174, when
speaking of the terrible effect, according to his opinion,
of *Tirailleurs* against a single column: " Let them yield
" the ground to the column as it advances; let them keep
" at a proper distance from it; let them fly about it,

dents, and particular circumstances, local and others, that in certain cases might modify the disadvantages with which his method appears to me to be attended. I shall therefore confine myself to saying that, generally, I think this method more dangerous than a resistance in front would be; provided, however, that that resistance be made according to principles calculated to obtain success: and I shall here venture some ideas on the dispositions to be adopted by an army, in order to repel an attack in columns.

In the first place, I would advance before the army a considerable body of *Ti-*

" trimming it well with an irregular fire, and its destruc-
" tion is certain." This would be an excellent manœuvre, no doubt, if the column advanced without being supported either by its own *Tirailleurs* or cavalry: but, on a contrary supposition, it would probably be secured by them from the enemy's *Tirailleurs*, whom they would prevent from approaching near enough to produce any great effect, or even reaching it at all.

railleurs, formed of light infantry, and trained to disperse and reunite rapidly. The effect I should expect from these *Tirailleurs* would be, to check those of the enemy; to prevent their approaching the army too near; reconnoitring its position; the disposition of the troops; the site of the batteries, &c.; and to discover, if possible, the dispositions of the enemy. For this purpose, I should order them to advance as far as possible, without endangering themselves, however, or ceasing to maintain a safe retreat. I would have them, likewise, supported by corps of light cavalry, which I would dispose in the most advantageous manner the ground would allow, and by some pieces of light artillery.

The first line of the army should be deployed; and, besides the cannon belonging to the regiments that compose that line,

I would erect batteries on the points which,
from the nature of the position, should appear to me the most favourable to the attack of the enemy, or the most advantageously situated to fire upon the flanks and
rear of his columns, as they advanced:
such, among others, would be the salient
parts of the position, the villages in front
of it, &c. It is understood, no doubt, that
I should proportion the force of those batteries to the importance of the points on
which they were placed, and that I should
have them properly supported by troops.
I would likewise support the first line by
corps of cavalry, forming a particular reserve for that line; and by some pieces of
artillery, which would serve, as I wanted
them, either to reinforce those of the first
line, or to replace such as might be da-
maged.

The second line should be formed into

columns at a hundred and fifty paces at most from the first, and supported like that by corps of cavalry, and by artillery in reserve. Lastly, I would place behind those two lines the general body of reserve, which I would compose for the most part of the kind of troops best adapted to the ground, regulating their strength according to local and other circumstances. I should also establish batteries in the rear of the army, to flank the passages by which it would retreat in case of a defeat; or I should at least previously reconnoitre the points most favourably situated for that purpose, so that cannon might be expeditiously transported thither, if it became necessary.

I suppose now that the columns of the enemy advance, covered, as it is natural to imagine, by their *Tirailleurs,* and supported by their cavalry: in the first place, if I required my *Tirailleurs* to dispute the ground inch by inch, they would undoubtedly direct their fire against those of the enemy's *Tirailleurs* opposite to them, instead of attending particularly to the columns which are my principal object. I think no Officer, who has seen service, will deny the truth of this assertion. In the second place, the fire of my *Tirailleurs*, were it even directed against the columns, would not be able to produce an effect equal to that of the artillery, which they would obstruct. In fact, supposing the columns, as I have reason to do, covered by their own *Tirailleurs*, who would keep mine at a distance, these would hardly be able to hit the columns, and only in certain directions; of course they could not injure them much. I should therefore order my *Tirailleurs* to fall back immediately upon the army, and to pass *a la debandade* to the rear of the

second line*; which they would execute
with the greater facility, as the men in the
first line would only have to turn a little
sideways, to let them pass. With respect to the cavalry attached to these
Tirailleurs, they should pass through the
intervals between the battalions, or some
files should open to make way for them.

---

* I here speak only of the *Tirailleurs* opposite to the
columns, and of those which, though stretching beyond
the space which the front of those columns occupy, are
not, however, at such a distance from it, that the fire of
artillery proceeding from that part of the line which they
cover cannot act with effect against them. With respect to the other *Tirailleurs*, I should wish them to fall
back as gradually as they could with safety ; because,
covering the portion of the line, before which they are
placed, they render it more difficult for the enemy to direct against it a false attack, and favour the execution of
the movements which, as I shall show, I should attempt
to make on the flanks of the columns where those *Tirailleurs* are already situated. When, however, after
retreating gradually as aforesaid, they found themselves
unable to keep any longer in front of the army, they
must, no doubt, like the other *Tirailleurs*, pass *à la debandade* behind the second line, and there form into a regular corps.

These movements, besides, being executed at a certain distance from the enemy, would be attended with no inconvenience.

The *Tirailleurs*, on coming to the rear of the second line, should immediately form into regular corps, in order to be ready to execute such movements as circumstances might afterwards require on their part. I should unite their cavalry either to that of the first line, or to that of the second, according as it should appear to me proper.

If the enemy's army were in a certain degree superior to mine, I might also (and it would furnish the means of strengthening the second line) dispose of the regiments or brigades which form the first line, in such a manner that they should leave between them intervals where the *Tirailleurs*, in retreating, should unite and form in two close ranks: but in this case, it would no doubt be necessary to take particular precautions that they make no kind of mistake in respect to their intervals, which would create a dangerous confusion.

As soon as the *Tirailleurs* had unmasked the line, the artillery should open its fire entirely against the columns, and without paying any attention whatever to their *Tirailleurs*. To this fire should be added, at the proper time, that of the infantry; and the attack should be resisted in this manner as long as it could be, observing to direct upon the columns the greatest number of cross fires possible. I say *possible*, because the infantry, at least those which are drawn up in a right line, cannot cross theirs but by firing obliquely, and because the oblique fire of musketry, to produce effect, should never, I think, proceed from a front much longer than that of a battalion*   I should likewise endeavour to bring against the outer flanks of the columns of attack, corps principally composed of cavalry, as far as the ground would permit.

If my first line were forced to fall back, it should retreat, and form behind the second, covered by its cavalry, which I should strengthen as much as I should judge necessary, from that of the second line: at the same instant, the second line should move forward in the charge step, to attack in columns those of the enemy, and I should immediately bring the general reserve nearer, in order to have it at hand,

* This principle appears to me the more applicable to the case before us, as I think that columns of attack should not exceed three battalions, with a front of one division. A greater depth would expose the parts most in the rear of the column to cross fires, without adding any thing to the force of its shock, and if the successive efforts of three battalions do not succeed, those of a greater number would not be more effectual.

and to be able to make use of it wherever its co-operation should become necessary *.

Whether, after having repulsed the co-

---

* The author, as we shall see in page 180, conceives a dreadful idea of an engagement between two columns, but in our days I do not think that it would be so bloody as he imagines: first, because, during the shock, only the heads of the columns would suffer, and that for a very short time; for the author himself acknowledges, that if two columns engage, they would be soon thrown out of their order, an assertion which I am bold to say is in favour of my dispositions; for, if there follows a breaking up of the columns, I shall certainly have rendered abortive the efforts of those of the enemy, which will then be obliged to engage in a different manner. Nor do I conceive that those columns, though broken up as columns, would, as he says, become a multitude murdering one another without order or principle; for it is probable that one would be broken before the other, and I think that the first in that state would retreat as speedily as possible under protection of the troops intended to cover it. I must, moreover, observe that, according to my dispositions, I gain on my side the probability of success, as I should oppose columns fresh and in good order against those of the enemy, which being fatigued and shaken by the successive actions which they have had to maintain, lie, no doubt, under great disadvantage.

lumns of the enemy, I chose to pursue him, or, those having succeeded in their attack, I should be obliged to retreat, local and other circumstances would, no doubt, determine the ulterior measures I should have to take. But, as those measures have no peculiar relation to the dispositions I here propose, I shall only say, that in case of pursuit, the part of my light infantry, which had fallen back to the rear of the second line, should return *a la debandade* to the front of the army, and throw themselves as *Tirailleurs* on the flanks and rear of the enemy, supported, as before, by corps of light cavalry and artillery.

Such are the general dispositions which I should think it would be in my power to make with success, against an attack in columns * : I shall now examine the prin-

* Some of these dispositions, such, for example, as the precaution of throwing a large body of *Tirailleurs* before

cipal advantages which they appear to me capable of giving.

First, By means of a considerable body of *Tirailleurs*, which I should push on in front of the army, I prevent those of the enemy, as I have already observed, from approaching and reconnoitring my position. By this, I render it impossible, or very difficult, for the enemy to discover my dispositions, the situation and force of my

the army, to establish batteries at the most important points of the line, and to reconnoitre at least those to which, in case of a defeat, cannon ought rapidly to be transported, in order to flank the passes through which the army is to retreat, are not peculiar to the case in which columns of attack are to be repulsed; for, whatever the order of battle may be, it is always equally advantageous to conceal from the enemy our own situation, arrangements, and the designs they indicate, while we endeavour to discover his. It is no less essential to take previously all the necessary measures to secure an army, if defeated, the means of retreating with safety. But, I have stated these dispositions, in the explanation I gave of my plan, in order the better to connect the parts of it, and to render it clearer in developing my ideas.

columns, my batteries*, &c.; and if I occupy a position strong by nature, or in-
trenched, he will not be able either to reconnoitre the feebler points, nor take the
prolongations that would be most favourable to him; nor, in short, determine the
direction of the capitals on which he would
advance with most safety. In a word, I
embarrass him in the fixing of his points of
attack, and in the manner of approaching
them with greater security; I keep him also
in the dark respecting those points whence
I may sally on the flanks of his columns;
and by that means I prevent his collecting
there, and at the points of attack, his cannon, which he is compelled to divide along
the whole extent of his front, in order to
be guarded every where; therefore his at-

* The enemy would then have no other means of reconnoitring my position, than that of advancing a strong
reconnoitring-party, which would throw more difficulties
in his way.

tack will be less powerfully supported by them, and, on the other hand, I shall come up on the flanks of his columns with greater facility. I prevent him further from withdrawing troops from some points, to strengthen his attack; for, should he withdraw troops, I should take advantage of the opening he gives me, to make a terrible attack on his flanks, at the same time that I resisted the columns in front; and the additional force he would by that means acquire at the points of attack, would probably be of little avail to him. Now, if he divides his troops, to be provided every where, he would not only be weaker at the points of attack, which would enable me of course to resist him the more easily, but my cavalry, unless his be greatly superior in number, would probably find means to overthrow him at some points *.

* I beg to refer the reader to the Commentary, page 109.

I suppose, however, that the enemy com-
pels my *Tirailleurs* to fall back, and that
his columns of attack advance: in the first
place, these columns will proceed under a
fire of artillery, which, on account of their
motion, the greater or less irregularity in
the direction of their march, &c., (I have
already expressed this opinion in my work
on Field Fortification) will be less destructive, perhaps, than some officers might
imagine, but which, nevertheless, if my
artillery is well worked, will not fail to
create a degree of irresolution in the columns on which it is concentered, and may
even occasion a commencement of disorder.
The columns will necessarily be more
shaken when they come nearer to me, as
they will be exposed to the grape-shot of
my artillery, and to my musketry. In
fine, if, in spite of all the obstacles I oppose to them, they succeed in breaking my

first line, this will doubtless increase their disorder, and in this situation they will be charged by the columns of my second line, supported by my cavalry; these troops, being fresh and in good order, will easily, I think, overthrow them. Add to this, that probably my dispositions will have enabled me to execute on the flanks of the columns the movements of which I spoke, and that those movements, very alarming to them, will prevent their advancing with the same security, in spite of the precautions which they might have taken to protect those flanks.

It will perhaps be objected, that if the enemy has also, as is natural to suppose, a second line, or a strong reserve, and I am attacked by these, after I had repulsed the columns, they must, according to my own principles, easily overthrow me; because, though opposed to victorious troops, they would come fresh and in good order upon them, when they must be necessarily shaken by the shock they had sustained. I could answer this objection in a variety of ways, which would all prove how ill-founded it is; but, I shall confine myself to observing, that while my columns were sustaining the shock, my first line, whose movements were covered, will certainly have had time to form again; I should then be able to bring it into action a second time in good order, and immediately follow up my advantages, so as to prevent, if I know how to conduct an army, all the offensive movements which an enemy should afterwards attempt. Besides, I have likewise my reserve: finally, if the objection made were admissible, it would be equally applicable to every disposition possible; for, whatever order of battle be adopted, it is evident that if an army, whose first

line has been beaten by the first line of the
enemy, renews the action by the means of
fresh troops, these, supposing the enemy
also to advance fresh ones afterwards,
would be subject to the like inconveniences that would be objected to me on this
head. These inconveniences then, admitting that they were to exist, would prove
nothing against my dispositions.

I shall observe, lastly, that the formation in columns of my second line, and
the great intervals between those columns,
render easy to me, not only the execution
of the movements of which I have spoken,
but every kind of dispositions which, according to circumstances, I should think
advantageous. Columns, besides, are easily moved, provided they are not too deep,
in which case it would be better to divide
them into a greater number: they may
be then transported rapidly to the points

where their service is required. If it be necessary to break the enemy by means of a deep body and a violent shock, the columns accomplish it; if, on the contrary, fire be wanted, they may be deployed. It is therefore a kind of formation adapted to all cases, and, in my opinion, the second line particularly should never be deployed till the moment it is rendered absolutely necessary by circumstances. It will be seen that I have dwelt more, in one of the commentaries, on the advantage of keeping an army as long as possible in columns.

I shall now conclude this Preface with some observations on the object and probable issue of an invasion of England by the French, supposing that they were to attempt it.

First, let me avow my disbelief of the possibility of the French landing a considerable force on our shores; such a force

as would render it necessary for this country to employ very extensive means in re-
pelling it. I am not indeed a naval officer,
and as my knowledge of that service is confined to general notions, it is possible that
I may not sufficiently estimate how far the
effect of the naval means which the French
seem disposed to employ, in conveying their
troops to our coast, may modify, as to them,
the principles hitherto admitted : but it is
on those principles, on the physical causes
from which they were partly established,
and on the nature of the element which
the invaders would have to cross, that I
found my opinion, which I support in the
following manner:

First, The French have not at their command a sufficient naval force to convoy
them safely ; they must therefore avoid
ours till they are disembarked. Secondly,
They cannot avoid them, but by the con-

currence of peculiar circumstances, which,
besides that they do not occur every day,
are more or less uncertain in their duration.
Consequently, admitting even that a favourable opportunity should present itself,
the French could not take advantage of
it, but by means of a great rapidity, not
only in their motions preparatory to embarking, but in their passage, and in their
disembarking: Now, how would this rapidity be reconciled with the inevitable
delays which, in a variety of respects, a
great expedition must occasion? How
would it be possible for such an expedition to leave its ports, sail the distance
which separates the enemy from us, and
land on our shores, where it would meet
a vigorous resistance, sufficient at least to
retard its disembarkation, without our
navy being able to come up in time to obstruct it? I can conceive, and we have seen

it happen, that, favoured by circumstances, a force, even a considerable one, may escape the most active vigilance, and get out of port; I can conceive too, that by means of the immensity of the ocean, and of the uncertainty of the course it takes, it may, before it can be overtaken, arrive and land at its place of destination, and especially when it meets no resistance in landing: but that it can meet with similar success in sight of a formidable navy, and with its destination known; nay, more, when the very points at which, or near to which it must necessarily be proposed to land, are, as I shall prove, in a manner determined, and would most certainly be defended, appears to me, I own, not only difficult, but impossible. It is in vain to say that the proximity of the two countries gives the enemy a fair chance of crossing rapidly; for, besides that the nearness

of the points, and the intelligence which
our ships of war would immediately receive of the motions of the invaders, might
enable the former to overtake the latter,
it is likewise to be observed, that a considerable force, encumbered by the train
necessary for it, and by a quantity of provisions of every kind, which it could not
do without, cannot be rapidly moved,
whatever means are used, nor landed in
the twinkling of an eye; especially when
it experiences a vigorous resistance at the
point where it attempts to disembark. If
it be urged that the enemy, in order to
lighten his march, to abridge delays, and
to divide our attention, may make his preparations at several points, I answer that
such a manœuvre would appear to me very
ill judged on his part, at least as to the
sequel of his operations; for, the consequence might be, that some of his divisions

e

might not arrive, or arrive too late; and, as he would then not have a sufficient force to act, not only would the object of the expedition fail, but the troops landed would be sacrificed. I must therefore continue to believe, that the enemy in his present situation, unless he be determined to make a desperate push, and stake all against all, cannot think of sending a considerable force to our shores, nor consequently of attacking us at home.

I shall suppose, however, that he will undertake it. In this case, he could only have in view one of two things; namely, to conquer England, which would require a war regularly conducted; or, to try a *coup-de-main*, the result of which, if it succeeded, would be very injurious to us. I shall examine these two propositions one after the other; and first, that of an invasion for the conquest of England.

I shall begin with observing, that the French must, in this case, necessarily bring with them magazines of every kind, and the more considerable the greater their number; for, as they would no doubt expect an obstinate resistance, and to struggle against powerful means of defence, they could the less dispense with this measure, that, after landing, they would have no communication secured with their country, so as to enable them to receive regular convoys. Even if the communication were open, the nature of the element over which the convoys would have to come would render their arrival uncertain. Nor could they, without the utmost imprudence, the consequences of which would be fatal to them, flatter themselves that they should immediately acquire an extent of country, sufficient to supply all their wants; not to mention that the country might be stripped of its resources. They would, therefore, have to secure means less doubtful than these, and to bring with them, as I have said, magazines of every kind, on pain of being compelled to surrender at discretion, even in spite of any advantage they might obtain at first, and which, from want of absolute necessaries, they would not be able to pursue. Now, these magazines, by increasing their embarrassment considerably, would render it still more difficult for them to act with the rapidity absolutely necessary in their situation. They would thus lose the advantage they might flatter themselves to obtain by numbers; or, if in order to avoid this dilemma, they were to diminish the force of their army, the extreme weakness which would be the result of this step, compared to our forces, would render their attempt more ridiculous than alarming.

This argument alone is enough to prove that the intention of the French, were they to attempt an invasion, could not be to conquer the kingdom; but I shall go further, and, foregoing for a while the interference of our gallant sailors in keeping the enemy from our shores, I shall suppose him not only landed, but that he has succeeded in obtaining an establishment, which he could not do without; and that, emboldened by this first success, he is going to advance into the country.

Now, without enumerating here the obstacles he must have had to overcome in effecting his landing, and establishing himself, which I shall consider afterwards, I answer for it that his first step has not been gained without a conflict; I answer for it too that his loss in this outset has not been inconsiderable. He must then, before he advances, secure the establishment I suppose him to have formed on the coast, and which would be to him a very material point. Of course he cannot avoid leaving there a number of troops proportioned to its great importance, which will so far weaken him in his offensive operations. But this is not all: he would be obliged to employ a considerable force in protecting his line of operation, which, daily threatened by numerous corps, would be in constant danger of being cut off Add to this, the immediate and great losses he must experience from the frequent, I might say the constant actions which he would have to sustain from the moment of his landing; for, even supposing that the English, resolving to act principally against his line of operation, which, as I have observed, is to him a very ticklish object and greatly exposed, should content themselves with keeping him in check in front, and

with occupying successively the positions
that would enable them to obstruct his
progress, he would be obliged to attack
them without intermission in those positions, as his situation would not suffer
him to rest; the delay of a week, of a
day, might ruin him : he must advance
rapidly; he has not a moment to lose in
spreading, and acquiring in the country resources capable of supplying the deficiency
of his own means, which, as he cannot
himself supply it, must be exhausted in
a given time, and that probably a short
one. He must, therefore, attack all that
opposes him, and endeavour to advance,
in spite of the irreparable losses he would
suffer: for, if he is compelled to stop he
is ruined.

What is to be inferred from these observations? Clearly, that before the enemy
has penetrated twenty leagues into the

country, supposing him to have been able
to advance so far, he will not have thirty
thousand men to bring into action at the
points where he means to act offensively,
even allowing his army to have originally
amounted to between fifty and sixty thousand, which, in my opinion, is improbable.
Is it with a force like this, a force too
that would be always decreasing as it advanced, that the enemy would pretend to
conquer a brave and numerous nation,
abundantly provided with all the requisites
of war, and who can easily and quickly
repair their losses of every kind; a nation
whom the very nature of their country
protects against the progress of an invading army, as the face of it in general presents an infinity of defensive means, and
as the central points whence troops would
be detached to the points of action, are
at so small a distance, that the forces would

be rapidly collected there, and particularly, as besides the goodness of the roads, the troops are provided with excellent horses, and carriages of every kind on the lightest construction; a nation, in short, who possess a moral energy the less likely to slacken from its being founded on the solid bases of the love of their king, of their country, and laws.

The conquest of England, or of any considerable part of its territory, would then, on the part of the French, be chimerical, which I would have demonstrated more at large, had I thought it necessary. They are as fully convinced of this truth as I am, and are certainly far from thinking of any thing like an enterprise of this nature. If, which I much doubt, they intend, if they can, to attack England on her own territory, their only object must be a *coup-de-main* against certain important points, and those are no other than our great naval and commercial establishments.

But, besides that those establishments are known, all are not of like importance; they do not all equally present the enemy, in case of success, advantages capable of indemnifying him for the expences of his expedition, or of counterbalancing the dangers to which the attempt would expose him. Add to this, that the distance at which he must land from the principal points, to have any hope of succeeding in his operation, is circumscribed; for, if he land far from those points, with the design of marching to them afterwards, he would involve himself in all the dilemmas into which I have demonstrated he would be thrown, were he to undertake the conquest of England. We see, then, that, notwithstanding the great extent of our coasts, the points of attack and of the enemy's landing are really determined, and that by leaving open to him, in some degree, all such as would lead to no material result, either on account of their great distance from the capital points, or because the country he would have to cross, to approach these, would render his march very difficult, we may without danger assemble more forces in positions central to several principal points, and near to them. By this disposition, we shall avoid weakening ourselves every where by too great a division, and secure the facility of sending speedily to the points of attack a sufficient force to prevent the enemy, if not from landing, at least from executing the *coup-de-main* he designs before there is time to march reinforcements to the troops first employed to guard the point attacked, and which, as I have said, would arrive rapidly from every quarter. The enemy, assailed on all sides by superior forces, which, after his landing, may safely be drawn from every part of the empire, must, undoubtedly, be soon overpowered, and an end put to his operations.

I have hitherto considered only the action of the troops: how many other obstacles would not the enemy have to surmount! Has not the wisdom of the government long ago foreseen the points where the empire might be vulnerable, and, seconded by the great talents of the commanders of the army, the known skill and abilities of the Artillery and Engineers, has he not taken all the precautions necessary to secure those points? How can the enemy conceive a probability of success, while those bays and roads, the possession of which is absolutely requisite to the execution of his projects, are defended by a formidable artillery, which, on his appear-

ing, would shower bombs and red-hot balls
upon him; while the mouths of the rivers,
and the entrances of the roads and bays,
are protected by fortifications, many of
which are so placed, as to be able to rake
all his vessels that should venture in; while
providing against the possibility of his
being able to land troops to attack the
batteries in the rear, care has been taken
to enclose the principal of them in forts,
or good redoubts, so that he could not
carry them instantly; while, lastly, the
principal landing places, at least those
that would be favourable to him, are
likewise furnished with good redoubts,
the support of which would give still
more effect to the action of the troops,
when, by means of a rapid march, rendered easy by their being near and by the
measures taken to convey them expeditiously, they arrived while the enemy was

landing, and probably before he had had time to disembark any considerable part of his artillery and cavalry.

The observations I have made warrant me in asserting, First, that the enemy cannot bombard by sea any of our great establishments, the acquisition or destruction of which he might have in view; Secondly, that to effect a landing would be extremely difficult, if not impossible. I suppose him, however, to have done it, and that he is advancing to execute the premeditated *coup-de-main*. He must, in the first place, surmount the obstacles he would meet at all the points which might enable him to approach the establishment, or bombard it by land; and this previous operation would necessarily require a certain time: he must then commence his attack upon the establishment itself, which would doubtless be so defended as not to be

carried immediately. Now, all these delays
would be favourable to the arrival of reinforcements; and the mass of troops would
increase in a proportion dreadful to the
enemy, who, as I have said, must necessarily be overpowered by it. I do not see,
then, that in any point of view the French
can succeed in an enterprise against England; and I boldly predict, that if they
should ever attempt an invasion, the result of it would be the destruction or
captivity of their troops.

ERRATA.

Page 113, line 7 of the note, *for* in forming columns, *read* in forming and deploying columns.

114, line 14 of the note, *for* in a level country, *read* in a level and open country.

THE

SPIRIT

OF THE

MODERN SYSTEM OF WAR.

PART THE FIRST.

A VIEW OF THE PRINCIPLE ON WHICH THE
MODERN SYSTEM OF WAR IS FOUNDED; OF
THE ALTERATIONS IT HAS PRODUCED IN THE
MILITARY SYSTEM ; AND OF THE RULES RE-
SULTING FROM IT FOR THE DIFFERENT CIR-
CUMSTANCES* AND OPERATIONS OF WAR.

CHAP. I.

*General Notions. Examination of Offensive Lines
of Operation; and First, of a single Line of
Operation, which, proceeding from a single Point
constituting its Base, advances into an Enemy's
Country.*

THE invention of gunpowder, and the consequent introduction of fire arms in military

---

* The Author appears to me to express himself too generally ; for if it be true, as I have observed in the Preface,

operations, rendered an immense quantity of ammunition necessary for the supply of those arms. It was gradually discovered that this new mode of destroying an enemy required a great display, and that an army was formidable in proportion to its number of lines of fire. In time this observation led to a change of tactics; and a system of spreading and displaying troops on a large scale, and of embracing a vast extent of ground, was adopted. In this new system it appeared that the number of soldiers produced the effect which formerly resulted from the personal qualities of the men; consequently, it became the object of the Powers of Europe to acquire the means of augmenting their troops: innumerable bodies were put into motion; and to the enormous equipage, which the ammunition drew after an army, was added the no less considerable one of the provisions necessary for so great a number of men and horses. The countries into which the armies marched, soon failing in the power to feed such multitudes, and to supply them with the means of fighting, it

that moral, physical, and political causes necessarily modify in practice the fundamental principles of the art of war, it follows that the author may be able to lay down rules applicable to the different operations of war, but not to the different circumstances, which are perpetually varying.

COMMENTATOR.

became requisite to think of magazines; and, in the end, the fate of armies was found to depend on the abundance of those places of resource, on their security, and on the facility of keeping up a communication with them: to establish them, and to fill them with stores, before the opening of a campaign, was then the object of attention. The positions best calculated to screen them from every insult, were previously and maturely considered; and, at the same time, it became a principle to manœuvre in such a manner as to cover them; not to go far from them but with great caution; and never to cease preserving with them those connections in which the strength of an army consists, and on which its success depends. Plans of campaigns were then formed *, fortified towns

---

* It is necessary, no doubt, previous to entering upon a war, either offensive or defensive, to settle a plan of campaigns; that is, to concert, in a general way, the measures to be taken to ensure the war's leading to the object proposed in making it, which ought to be calculated according to the data supplied, not only by the situation, local circumstances, &c. of the frontiers of the Belligerent Powers, but also by the comparative view of their forces, and resources of every kind, and by the nature of the war to be carried on. These measures may be considered under two heads, political and military.

marked out as fundamental points; in a word, a *base* was fixed where magazines were established,

To the political are referred, the alliances to be formed, the treaties of subsidies to be entered into, the discussion of the causes that render the war necessary, and consequently that of the grounds on which the treaty of peace that is to terminate it should be founded, supposing the success of the war to be equal to the hopes entertained of it : to this head too is to be referred, the advantage to be taken of enmities, of factions, of civil dissensions, of private ambition, of the jealousy conceived by states, and even by individuals, of one another, and of the differences arising in the same nation, from customs, religious tenets, and opinions of every kind : lastly, it belongs to political measures to foresee that certain governments, neutral at the beginning of the war, will afterwards become hostile; and to form beforehand provisionary plans by means of which their aggressions may either be resisted if they declare themselves, or anticipated by an attack before they have completed their preparations.

Then follow the military measures, which are to be so concerted as to contribute the most effectually to ensure the success of the war: the first is the determination of the points on which the operations are to begin. If it be intended to act *offensively*, it is necessary to begin with examining what part of the enemy's frontiers, by its situation, the nature of the country, and the concurrence of other circumstances, which I shall hereafter more fully explain, not only furnishes ampler means and greater facility to attack him there with advantage, but will likewise lead more directly, after it is mastered, to the principal object proposed to be attained : against that part should the principal opera-

whence lines of operation issued, and of which the object was to protect retreats, as well as to favour attacks.

tions be directed. If any other part of the frontiers be found favourable to a powerful diversion, and which should likewise offer the advantage either of leading the enemy into an error respecting the point where the principal operations are projected, or of forcing him to weaken himself at this point, to defend that against which the diversion is to be made, it will be requisite also to march thither, provided, however, that the army be not thereby exposed to any of the inconveniences planned for the enemy; as, for example, that of being weakened by too many divisions. It is also to be observed, that, as it is possible that a war begun offensively, may decline into a defensive one, and that the enemy may himself attempt diversions, it is necessary that such of the frontiers of the offensive army as might present him an advantage, and particularly those containing the magazines, or which would enable him to take in flank or rear the positions of the army, while advanced, should be completely covered. In short, for every hypothesis, provisional plans should be settled, in order to be prepared against whatever may happen.

When once the points of attack are determined upon, the base is settled whence the operations are to proceed, and on which are to be established the magazines, which are, as much as possible, to be so disposed as to divide the attention of the enemy, by drawing it to several points at once. Demonstrations are made to mislead him, and if he is deceived, advantage is taken of it. The direction which the lines of operation are to follow, in order to terminate at the first intended point of action is then determined.

The utility of this base, its figure, and its dimensions, objects of such importance in the

The safer plan is, to choose, not only the shortest and easiest way for them, but likewise that which affords the enemy the fewest means to molest them. There are countries in which this direction is self-traced; flat and open countries are generally of this kind: but there are others, which being woody and mountainous, present the enemy an infinity of means to obstruct the communications and attack the convoys, and which, consequently, require the greatest precautions. In such countries, to secure the conveyances of the supplies, and render them easy, it is necessary, when the army is on its march, that it should be preceded by a great number of pioneers, in order to remove, as much as possible, every kind of obstacle likely to obstruct or endanger the communications; to destroy the roads and tracts through which the enemy may steal upon them, and to place, at'certain distances, posts more or less numerous, and more or less strongly entrenched, according to the nature and importance of the points where they are situated, to serve as so many steps, by means of which the supplies may be conveyed with safety.

Nay, sometimes, when it is foreseen that a war will be of long duration, exertions are made to conquer, in some sort, nature itself, by blowing up rocks, cutting roads through forest, and carrying some of them to the points most proper for the establishment of the depots of provisions and ammunition that are to go along with the army. If there be a navigable river at hand, as many boats as possible are collected upon it, to take advantage, of the conveyance by water, which, besides a considerable reduction of expence, may be the means of the operations

present system of war, particularly demand our attention, and shall be principal subjects for our consideration.

proposed being executed with greater facility, and at an earlier period ; for there are countries which can be entered only at particular times, on account of the forage, or of the state of the roads. Sometimes, too, rivers not actually navigable, may, at a proper time, be rendered such at a small expence ; and when that opportunity offers itself, advantage must no doubt be taken of it.

Attention is also paid to the storing of the magazines; the number and stations of the troops intended for the protection of the frontiers are settled ; the fortresses that require it are repaired, particularly those which in the course of the campaign, or of the war, may be exposed to attacks ; and they are, more or less, furnished with stores and provisions, according to the extent and proximity of the danger they may run ; and, lastly, according to the nature of; the country in which the operations are to be carried on, to the number, quality, and kind of troops that the enemy may bring into the field, the strength and composition of the army, as well as the number of cannon and pontons to attend a, are regulated : the general officers to be employed in it are' appointed, and the provisional instructions necessary dell-' vered to them; the different corps of which it is to be formed are drawn nearer, and when the time fixed for the commencement of the operations arrives, they are assembled, care being taken to plan as well as possible their junction, in such a manner as to meet no obstruction from' the enemy, and to mislead him in respect to the point towards which they are to be marched, in order, by this means, to anticipate him there. Except in a case of abso-

The necessity of attending so particularly to the base of the lines of operation, made

lute necessity, the army should not be assembled at a great distance from the first point of action ; for it is a rule, in an offensive war, that the first movements particularly, should be extremely rapid, and if the men and horses, from a state of rest, have immediately to make a long march, they may be so much fatigued as not to be able to act afterwards with the necessary celerity.

I have said nothing of the levying of new troops, of the augmentation of the old ones, of the purchase of horses for the use of the cavalry and artillery, for the conveyance of provisions, and for the service of the army in general ; nor of the construction of wains, carriages, ordnance-stocks, covered waggons, &c. of which, no doubt, the army will stand in need. It is generally known that these preparatory measures are indispensable before the commencement of war ; it is also known what influence secrecy has on the success of military operations in general.

At length the army advances towards the first point of action, and then it is that combinations rendered necessary by a multitude of local and other circumstances, of which there is an endless variety, determine the movements and operations which it is to execute. Has it been able to anticipate the enemy in the field ? Will it be able to come up with him before he has had time to collect his forces from their quarters? or are they assembled, and the enemy already on the defensive ? Is this defensive strictly so, or will the situation and resources of the enemy allow him to wage a defensive-offensive war ? Is the country attacked flat and open, or mountainous and entangled ? Is it protected entirely or in part, either by a position or by fortresses ? If

it also necessary to determine accurately the point to which those lines were to be carried.

by a position, what is the nature of it, and what are the means of taking it before the enemy, if he does not yet occupy it, or of dislodging him if he does? How are the fortresses situated? Do they stand upon the extremities of the frontier or at some distance from it? And in the latter case, is it not possible by manœuvring to advance a column between the enemy and them so as to cut him off? How is he to be driven behind his fortresses? Is it by vigorously following up the first advantages the army may have obtained, to prevent his rallying; or by well combined manœuvres, or in fine by a battle?

Further, what is the strength of the fortresses that cover the enemy's country, the ability of the governors, the number and goodness of the troops appointed to defend them? Are these fortresses in good repair, sufficiently garrisoned and provisioned in every respect? Which are those that on account of their importance (and this importance should be estimated less from their internal strength than from their local situation, and the kind of obstruction or of support they may give in respect to the operations) must necessarily be besieged? What is the probable duration of the resistance they will make? Which, on the other hand, are the fortresses that may be neglected, or merely masked? In fine, which are the fortresses that it would be of importance to keep as such after they are taken; and which those it would be better to demolish, to avoid being weakened by the garrisons they would require, and that without yielding any advantage capable of counterbalancing this inconvenience?

The general and vague object of conquering the
enemy, and of driving him as far as possible, was

If the object be to pierce the enemy's line of defence, or
to dislodge him from a country he occupies, and which he
can defend only by manœuvres, what means are to be employed to accomplish it? Shall it be, by so confining him in
his forage as to compel him to abandon such a point as may
be desired? Shall it be by menacing at once several points
of his frontier, so that when he has weakened himself by
sending detachments to cover them, he may be attacked to
advantage, and those very detachments be perhaps cut off?
or, shall it be by manœuvring before him so as to excite his
apprehension for some important point, and to keep him
thus in check, till a movement is made on his flank, that
shall oblige him to fall back for fear his communications be
cut off? Lastly, Shall a diversion, real or feigned, be attempted, or a decisive battle fought at the very opening of
the campaign?

Suppose that the country invaded be flat and open, and
without any fortresses to defend it; let us even suppose
that the enemy himself has voluntarily abandoned it, after
having deprived it of all the resources it might have
presented to the offensive army; how is that army to establish itself there firmly, and in such a manner as to secure
its depots, communications, &c. should the enemy, taking
advantage of the extension which the occupation of this
country would require, or with the help of reinforcements
he expects, afterwards act offensively? What might he then
undertake; where direct his efforts with the greatest advantage to himself? What are the points to be fortified,
and to what degree should they be so in regard to their

abandoned for that of conquering him in a particular point, driving him from a particular posi-

local situation, and their respective importance? How are
the winter-quarters of the army to be secured? or the
offensive operations for pushing forward to be prepared?

It is not enough to have arranged the general measures of
which I have been speaking; it is necessary, besides, to consider the particular means of execution, which an infinity of
local and other circumstances must regulate; the choice of
those means, and the manner of employing them.

As to acting *defensively*, the chief thing is to endeavour,
in the first place, to penetrate into the designs of the enemy
relative to the first principal point of action, and to those
where it may be in his power to undertake some important
diversion. The knowledge of one's own strength, resources,
and frontiers, and sometimes of the manner in which the
enemy disposes his magazines, become so many guides to
lead us to the attainment of this momentous object. It is
also to be examined which of the frontiers of the enemy are
exposed to an advantageous diversion, supposing there were
means to attempt it; and what territory it would be better
to leave open than endeavour to defend, either because the
enemy's advantages in occupying it would be less to him
than the disadvantage of dividing into parties to cover it
would be to the defensive army, or because it is foreseen
that he will not for want of means be able to maintain and
establish himself there; or, lastly, because, at a future period, it will be possible to dislodge him. In consequence of
these data, magazines are established and lines of operation
settled, the fortified places are put into repair, provisioned,
and furnished with a number of troops sufficient for their

tion, pursuing him to another, and stopping·
judiciously in the midst of triumphs, not so much·

defence. And here I must observe, that every place situated in such a manner·as to favour the penetrating into an
enemy's country is a capital point, and cannot be too carefully attended to. Political and other measures, such as
the levying of new troops, the augmentation of the old, &c.
which may contribute most effectually to foil the enemy's
projects, are at the same time taken into consideration.

After this, follows the settling of the camps, of the positions to be occupied, and of the passes and openings to be
guarded, to oppose at the outset the first approaches of the
enemy, and afterwards his further progress, if, unhappily,
the defensive army is compelled to fall back successively
from position to position. Every advantage is taken of the
ground, to which are added the resources of art in entrenching the points that are judged to be of sufficient importance. It is not enough to be prepared against particular
movements intended by the enemy, but all that may be
undertaken are to be forseen, and for every case plans are to
be preconcerted to resist him every where; either by opposing him in front, if the nature of the position occupied
admit it, or by falling on his flanks, and menacing his lines
of operation, if the thing be practicable, or by both these·
ways.

At last, these precautions being all taken, and the probable moment of invasion at hand, the Commander-in-
Chief assembles the main body of the army in a central
position, whence he may move rapidly to such points as
will enable him to oppose the more effectually all movements
which the enemy may attempt to make against the line

on an estimate respecting the enemy, over whom a superiority might be still maintained, as re-

of defence. Detachments, as the author will inform us, ought not to be inconsiderately multiplied, that the army may not be divided, without an absolute necessity. Nor can the means of anticipating the enemy in the field be too soon secured; for if a defensive war be of itself very dangerous, how much more so must it be when the troops are suffered to be surprised in their quarters?

Defensive operations may be considered in different points of view. In the first place, they may be foreseen and agreed upon, either on account of a great disproportion of strength, and of the paucity of resources the country offers to act offensively against an enemy greatly superior in number, or in the kind of troops suited to the country; or, because defensive arrangements, on a certain point, suffice to support the movements of an army acting offensively on another point; or, in short, because as the result of these movements must influence the subsequent conduct, the defensive is observed till that result is decided. An army may likewise, for a time, be reduced to the defensive, either to wait for reinforcements, or for the departure of such detachments as the enemy will be obliged to send off, at a certain period, and which will weaken him; or because it is foreseen that some reasons will compel him soon to evacuate the country given up to him, without its being necessary to attack him there. In short, defensive dispositions may be unexpectedly rendered necessary by the loss of a battle, by the unsuccessful issue of an offensive operation, and by a diversion which the enemy may have made. It also happens that, in spite of a kind of superiority which the enemy may have, the army is not reduced to a strict defensive, but, according

specting the victorious army, in order that
its forces might not be exhausted.   All the

to circumstances, may act offensively against him.   This
kind of defensive war is no doubt the most advantageous
that can be waged, and though reasons of the greatest
weight should at first oppose the entering upon it, nothing
should be neglected to acquire the means as soon as
possible.

It will be easily conceived that the different circumstances
I have been stating will necessarily induce some alterations,
not only in the plan of the campaign, but also in the conduct of the war. If, for instance, the army be restrained to a
strict defensive, on account of its great inferiority, a general engagement, the issue of which may compromise the
country and the troops, should be avoided as much as possible. The plan should be to destroy the enemy by skir-
mishes, by attacking him on his marches, by falling on his
detachments, by enterprises on his depots and lines of
operation, by waiting for him at the passage of rivers and
defiles; and if the army offer battle, it should be only after he
is skilfully drawn to a ground which not only renders his
superiority of no avail to himself, but also affords the defensive army such decided advantages, that without having
to fear the issue, it must, in consequence, be in a manner
sure of giving the enemy a considerable check. If, on the
contrary, the offer of battle be made by the enemy, it should
not be accepted till it is absolutely unavoidable, which almost always supposes a series of anterior movements ill combined; for an army should always manœuvre in such a
manner as not to be compelled to fight, unless it wishes it.

But in an offensive defensive war, where actual means
and future resources allow a greater latitude, move-

operations of war, then, are now divided into three principal parts; namely, the *base of*

ments may be made and measures adopted without danger, which, in the state of a strict defensive, would be extremely imprudent, and could produce no advantageous consequences. It is possible, then, to form different combinations. May not the enemy be anticipated at the opening of the campaign, and be attacked himself before he has time to collect the troops from the quarters ? These quarters being spread over a certain extent of territory, it is not possible in spite of the enemy's superiority, by marching to a limited number of points, to bring against him a greater force than he has at each of those points, and thus oblige him to fall back to unite his troops ? Which are the points most favourable to the success of the operation ? How are the magazines of the enemy protected ? Is it possible to seize and destroy them, or are they placed in fortified towns, which secure them from a coup-de-main ? And if the latter be the case, will there be time to take those towns before the enemy can relieve them ?

Suppose the magazines taken ; is the expedition to end there, or is it possible, without fear of losing the fruits of it, to improve the advantage of the first success and drive the enemy back from the position to which he may have retired ? What is the nature of that position, and what means have the assailants to compel him to abandon it ? What means has he to force them to retreat, and to stop their movements? Have they no reason to fear some enterprise against their lines of operation, a diversion, a battle, of which the issue may be disadvantageous to them, &c. ?

Let us now suppose, that defensive operations are to be adopted at one point, while offensive ones are to be carried

operation; the *line of operation,* and the *ob-ject.*

on at another. How is the defensive side to be concealed from the enemy so as to prevent him from disconcerting the general plan of the campaign? For if he suspects that side, he will not fail to march thither a superior force, which will render it necessary to draw off some of the troops from the point of offensive operations ; so that besides the disadvantage of being weakened at that point, where perhaps nothing could be afterwards undertaken, there would be danger that the reinforcements dispatched to the defensive side would not arrive there in time. This would be losing on all hands. Besides, is the country favour-able to this kind of diversion? Is it close, intersected by defiles, full of impediments? Is there in it any river difficult to pass by which the enemy may be obstructed? Is there, on that river, a fortress which the enemy must besiege before he can advance further, and which furnishes the means of coming up in time to stop him, &c. ?

Lastly, suppose that an unforeseen defensive plan is the unfortunate result of a battle lost, a fortress taken, quarters surprised, or the failure of an offensive operation, &c. How is the ground to be disputed with the enemy inch by inch, and how is he to be prevented from using his advantages, without risking a general action, which, in such a situation, would probably be attended with total ruin ? How is he to be prevented from firmly establishing himself in the country which it is necessary to abandon to him ? How, in fine, is the renewal of offensive operations to be ef-fected?

I have stated briefly the bases on which plans of cam-paigns, whether offensive or defensive, are laid down, and

As the only object of establishing a base is to have fixed magazines, and as the fortresses where the magazines are placed· constitute the military base, whence every operation, to produce effect, must necessarily proceed, it follows, that when an army is encamped very near a principal magazine, there is no line of operation; for, an army, in that position, is in complete security, is at the fountain of its existence, is sustained without convoys, and consequently has no occasion to manœuvre in order to protect its provisions against the attacks of the enemy. Now, the lines of operation commence when an army advances so as to leave its magazines behind it; for, properly speaking, it is the convoys that form these lines; and the reason for tracing or combining them beforehand is to secure the convoys; but when convoys are not

enumerated the most general combinations, according to which they are conducted : there is, no doubt, an infinite number of particular considerations, arising from local and other circumstances, which cannot be all foreseen, or particularized, even in a work the most detailed, and which theory will never be able to point out. The first principles furnished by theory begin the education of an officer ; an observing mind, activity, genius, *coup-d'œil,* and the experience acquired by a few campaigns, complete it.

COMMENTATOR.

C

wanted, the lines of operation disappear and are lost in the base.

Hence it appears that lines of operation are always directed forward, against the enemy's country. I say against the enemy's country, and not against the enemy himself; for, the object of war at present should much rather be those places which contain the means of an adversary's military power, than men. Thus, to march forward, in respect to the operations, is not always to march in the direction of the appearance of the soldiers, that is to say, straight to the points towards which the heads of the columns are turned; but to those where the judicious appearance of the army, that is to say the soul of him who leads it, shows it to be destined.

Lines of operation, then, according to the acceptation of that term, are of an offensive na_ ture. In explaining, by means of the figures annexed to this treatise, the mode of manœuvring in order to penetrate into the heart of an enemy's country, an idea of the real character of the modern system of war will be better obtained than by any other method. Let us first consider the case of a line of operation issuing from a single point, constituting its base, and directed towards a certain object, by advancing into an enemy's country.

Let A (*fig.* 1.) be the base, B the object, and C the army, proceeding on its march from the base towards the object. It is evident, that if the enemy D, opposed to the line of operation A B, advances on the rear of the army C, making but a slight attack on the detachment E, this attack is a feint to keep that detachment in its position, and that the army C will be thereby compelled not only to stop its march, but likewise to change its offensive into defensive operations, contrary to its original design *; for the line of operation A B being the only way by which its convoys can approach it, it is of the

---

* This is true; yet some elucidation may not be amiss to render the demonstration clearer and more striking. The army C, which has advanced towards B, has left the corps E in its rear, to protect it, and to defend the line of operations A B against the enterprises of D; now, if D advance towards the rear of C, making but a slight attack on E, it is evident that this attack must be a feint, and that its object is not to dislodge E, but to keep it in its position, and by that means to prevent it from opposing a movement that D intends to make against some other point of the line of operation A B, which he designs to cut off; and as it is of the utmost importance that C should secure that line, which is the only one by which its convoys can reach it, there is no doubt that it will be then obliged, as the author says, not only to stop its march, but to change its offensive into defensive operations, contrary to the end proposed.

COMMENTATOR.

C 2

utmost importance that that line should not be
cut off by the enemy D; because having no longer a passage for its supplies, and convoys de-
spatched from their own magazines being, as we
have said, the only means of supporting modern
armies, it would follow, in the case before us,
that the situation of the army C would be like
that of a man who, having staked his whole fortune on a hazardous enterprise, would have reason to tremble at the slightest miscarriage. An
army's lines of operation may be compared to
the muscles of a man's body, on which the motion of the members depend: when the whole
moving spring of a member is confined to a single
muscle, the loss of which would render it useless,
it is the more important to defend it from
every hurt; so, a single offensive line is, to an
army marching towards an object, a part singu-
larly sensible, and cannot be too carefully guarded from contact with the enemy.

If the army C had previously carried the important post B, the object of the operation, and
the detachment E was attacked or threatened,
the general would be under the necessity of
detaching troops to the assistance of that corps,
in order to protect the line A B and to keep
up his communication with A; but in doing
this he would weaken the army at B, in such

a manner that it would probably be unable itself to resist a direct attack. If, supposing B not carried, the army had commenced a siege against it, it would be impossible to continue it from want of sufficient force: or, if B was only an entrenched post, from which it was necessary to drive the enemy, the army C would not dare to hazard the attack. By this, and by what was said before on the supposition that C had not arrived at the object B, but was only on its march towards it, we see how detrimental this diversion of the enemy against the line of operation is, and that it necessarily puts an end to all further offensive operations on the part of the army C.

Let us now suppose that the army C, having carried the post B, is firmly determined to hold it, without suffering itself to be shaken by any of the movements which the enemy may make against its line of operation A B; it will be in danger of seeing its communication cut off, not only with A, but likewise with the detachment E; for the space between E and C must necessarily be so great as to suffer the enemy to march a detachment into the interval, for the purpose of acting, not only on the rear of C, but on the flanks of E, so as to compel the latter to retire. It may be objected that the detachment of the enemy D may

be beaten by E, in which case the line of operation A B will be secured, and the flanks as
well as the rear of the army be free and without
apprehension: but here things are not on an
equal footing. If E is beaten, the army C is
lost; if, on the contrary, the enemy is repulsed
by E, his own country is open to him, and his
retreat and supplies are always secure; for in
F, G, H, I, he has magazines which can never be
cut off by the army C. This army may possibly
attack and take from the enemy a magazine at
K; but the others, especially G, H, I, are at too
great a distance in the rear of D for C to be able
to reach them. But it will be asked, what if
the enemy D have no magazines at those points?
I can only answer, that it would be his own fault,
and that he is not ready to make war: I argue
upon things only as they ought to be, that is to
say, taking for granted that they are conformable
to the dictates of sound reason. I repeat, therefore,
that the enemy D, in his operations against the
rear of C, is always secure of his communications with his magazines; the greater number of
troops, then, he detaches against AB, or the more
he advances himself, the more must C weaken
itself in respect to the object B, in stretching
back, not to be separated from A. Besides, the
enemy D may also make a diversion on the point

L, in the country of the army C; or it may even attack A, if it be weak enough to be carried rapidly: for, as to a regular siege, C would soon fall back to save A. The line of operation A B is furthermore open to attacks on the side of M; for the enemy has it always in his power to assemble forces in his own country. On this supposition, the army C will be obliged to station a corps N on that side, in order to oppose the enterprises of M, which will weaken it still more at the object B, and to take the defensive figure of an oblong square, instead of offensive operations, which it had in view. We see, then, that in all cases it is extremely disadvantageous to act on a single line of operation, which differs little from the right line.

It is difficult to determine, in a precise manner, to what distance an army may remove from its magazines, when penetrating into an enemy's country, and at the same time avoid the disadvantages we have been explaining, for it depends upon a great variety of circumstances*. I may, however, venture here to lay it down as a prin-

---

* There is no doubt that circumstances, which are of endless variety, must, in a great measure, determine the distance to which an army may, without danger, remove from its magazines, when advancing into an enemy's coun.

ciple, that it should not be more than three days'
march, and that within that compass there is no
great danger, as the army by a single retrograde movement covers its rear and flanks, and
protects its convoys. But I must, at the same
time, observe, that in this case, the advantage
of an offensive operation would be very trifling,
even where a fortress of the first order was the
object of attack; and the attacking army would

try; and for this reason it is impossible to fix that distance
invariably. It may, however, be laid down as a general
rule that an army should in every case provide for the certainty of receiving its convoys in time, and in a regular
manner: especially when the country in which it advances
does not furnish it with resources that might serve for a
little while, instead of the supplies from the magazines;
for, besides the inconvenience of wanting provisions, it may
also happen from this circumstance, that the army, being
obliged to wait for them, shall be rendered incapable of
making movements, not merely advantageous, but perhaps of the first importance. The necessity of securing, in
all cases, the magazines on the frontiers, as well as the lines
of operation, has already been stated; I shall further add,
this ought to be most particularly attended to, especially if
it has not been possible to take the field before the enemy;
for, having had time to assemble his forces and to prepare
for the defensive, he may with less difficulty act against the
supplies of the offensive army, and even perhaps by that
means cause it to lose a whole campaign by obliging it to
take the defensive.—COMMENTATOR.

perpetually have to dread diversions in its own country, and on its flanks, as the enemy would be near his own magazines. If the belligerents have both a range of fortresses, or military base, to cover their respective countries, things are then on an equal footing. Each may besiege a fortress of the other at the same time, and the point will then be who shall be most expeditious *.

* These assertions of the author's appear to me to admit of an argument. There can be no doubt that lines of operation should lead to an object, the possession of which is of essential importance; otherwise the success, however complete, would be useless: but that the proximity of that object, as the author says, renders the advantage of an offensive operation directed against it of little importance, is what in my opinion requires examination. Let us suppose, in fact, that the enemy, having skilfully taken advantage of the nature of his frontier, has built on one of its points a fortress, by means of which he may easily penetrate into his adversary's country; or that its situation enables it to cover, either other places or a great extent of territory; can it be said that the advantage of an offensive operation against this fortress, even supposing it to be very near the frontiers of the attacking army, is of little importance? What imports it that the first point of action is very near the base, if by taking that point great advantages are gained for the subsequent operations? Nay, what imports it that the siege of a fortress, situated at that point, stops for a time the progress of the offensive army, if, when taken, that army shall be able, while it ensures its operations, to compel with ease the enemy to

I now resume my observations relative to a single line of operation. It may be said: if a country have on its frontiers a range of fortresses which consti-

evacuate a great extent of territory, and to open an entrance into his country, where that very fortress will the more enable it to establish itself firmly. It is not then, because the first point of action may be very near to the base, that we should deem of little importance the advantage of an offensive operation directed against it; and in our estimate of the degree of that advantage, we should be led by a consideration of the effect which the possession of that point may have on the issue of the subsequent operations.

I grant that if, independently of that fortress, the enemy has others near, with magazines, the proximity of those magazines would favour a diversion he might be inclined to make on the flanks of the offensive army, or in its country, against which it would be obliged of course to take measures of security; but these measures would be equally indispensable, should the first point of action be at a greater distance from the base : for, the first rule to be attended to before you invade an adversary's country, is to take care that he can do nothing successfully against your own. The only question then is, whether, in the hypothesis advanced by the author, it becomes more difficult for the offensive army to protect itself against diversions than in other cases. I do not think it : for, if, on the one hand, the nearness of the enemy's magazines favours him in an operation of this kind, on the other the offensive army may with less difficulty protect its flanks and rear, and consequently its lines of operation, for it is nearer to the fortresses which constitute its base; and the author himself allows that its danger is not very great. Besides, if these fortresses are within reach of one another, the offensive army will be near the troops which it would of

tute a military base, an army would be safe from all danger, even in acting on a single line of ope-

course have previously detached to secure it from the diversions of the enemy, and will therefore be able, as occasion requires, to support those detachments, or be supported by them.

I now come to the following assertion of the author's: " If the belligerents have both a range of fortresses, or " military base, things are then on an equal footing. Each " may besiege a fortress of the other at the same time, and " the point will then be who shall be most expeditious." I take upon me to say that this proposition also requires explanation. In the first place, I have already taken notice of the means which the offensive army has to oppose the diversions of the enemy, and it may be presumed, that it will not be impossible for it to prevent a siege which he might intend to undertake : but, supposing it should, are the fortresses that constitute the two bases equally important, in their respective positions, either in the nature of their works or in their situation ? Will such a fortress, which the enemy might be able to take, give him as great advantages as such another which may be taken by the offensive army, will yield to the latter ? Will the resistance made by each of them be equally long, and may not the offensive army, though unable to prevent the enemy from besieging one of its fortresses, which it is of importance to preserve, foresee that it will succeed in its operation before the latter has time to carry that fortress, and then compel him to raise the siege either by giving him battle, or by means of a movement which the possession of the point it has gained will enable it to make on his flanks or rear ? Or, if the enemy can make himself master of the fortress; will not that army be able, after succeeding in the object of its operation, to attack him immediately,

ration; for besides that the garrisons of those places would menace the rear of the enemy if he advanced far, new lines of operation might be thence projected, still securing a retreat, in case it should be necessary, to the fortresses. On

and retake the fortress, before he has time to repair the damage it sustained during the siege, and to lay in fresh supplies, the magazines of the garrison being in all probability either entirely or almost exhausted? Hence we see, that, although the belligerents should have both a range of fortresses to cover their respective countries, things might not be, as the author pretends, on an equal footing, and that many different combinations may vary the general principle he has laid down, though that principle be in some points well founded.

I shall here take notice, that the author has partly founded his Strategic demonstrations on the two following principles: first, that in our times, it is only the great Powers that go to war, and that consequently there cannot be a great disproportion of force between their armies : secondly, that the division of the forces of which an army is composed becomes injurious to it, as it weakens it, and may prevent its attaining the end of its operations. I am far from intending to dispute the truth of these principles in a general point of view, and I shall only observe that, in consequence of the difference between the political, moral, and physical constitution of different countries, it frequently happens that States, though really equal in power, cannot, at the beginning of a war, send an equal number of men into the field, or maintain them constantly on the same footing during the whole of the campaign. We may therefore be allowed to suppose that at one of those

this hypothesis an enterprise would no doubt have a sufficient base, for, in fact, the army would have in its rear several lines of operation for the passage of convoys, though they were not actually used, and though a single magazine sufficed for the supplies. But this is not the question. By acting on a single line of operation, I mean, either when there are no magazines within reach of a single point A (*fig.* 1.) which constitutes the base; or when an army is separated from the fortresses, where those magazines are kept, by any of the enemy's fortresses, which may prevent a communication with them : thus, for instance, the fortress L

periods, one of the two armies acting against each other, may find itself in strength and means of every kind superior to that opposed to it, and avail itself of the advantage to adopt measures which, without that advantage, might endanger it. From this, and from local and other circumstances, arises the possibility of modifying in practice most of the fundamental principles laid down by the author, though those principles are extremely just in the point of view in which he considers them. He has certainly done much in laying them down, and as to the variations to which they are liable, genius and experience alone can point them out.

I request the reader's attention to the observations I have made in the last paragraph, as he will find them of use in the perusal of this work, and I do not mean to return to the subject of them.—COMMENTATOR.

(*fig*.1.) belonging to the enemy D, renders the magazines that may be in the fortress O of no use to the army C; because the former obstructs the line of operation which C might have intended to open from the latter: and it may be laid down as a general rule, that those magazines which an army finds separated from it by hostile fortresses should be considered by it as non-existent. Now, it is most probable that the enemy will not fail to erect fortresses opposite to all those before him, just as a body of troops is always opposed to each of those of the adverse power, to keep it in check.

It will be necessary, then, in order to establish a sufficient base for the line of operation A B, to have, independently of A, other fortresses situated at a proper distance from that point, and on a line with it. This rule holds equally good, whether A be a fortress without a correspondent one in the enemy's country, and which for that reason the army C makes use of for its point of departure, or whether it be a fortress taken, at the commencement of the war, from the enemy D.

It was incumbent upon me to inquire minutely into all the disadvantages of a single line of operation, because these inquiries compose the foundation of the edifice I mean to erect in the following chapters. If the principles I have

laid down are granted, I am certain that the
consequences of them will not be denied.   It
was the more important that I should expose
these disadvantages, that the generals of our
times, and the most celebrated too, seem by
their conduct to think nothing of them ; and
that they are much more intent on pushing forward as far as possible, than attentive to diversions*, which I mean to show in the third part
of this work.

* It is undoubtedly a great fault to expose oneself to the
important diversions which the enemy may make, for the
purpose of pushing forward more promptly. Such was that
committed in the Low Countries in the year 1794, and
which was the cause of the failure of the campaign.   If the
allies had begun with the siege of Maubeuge, the fall of
which would have ensured that of Philippeville, and covered
the country, the French would not have been able to make
any diversion towards the Sambre.

Diversions well combined become, generally speaking,
a very powerful means of success, not only in defensive but
even in offensive war. I shall here observe, that by well
combined diversions I only mean those which have an important object in view ; as, for example, to invade a country
which it is of importance to the enemy to preserve, to besiege a town which it behoves him to defend, to carry off
his magazines, to cut off his lines of operation, &c.   Diversions of which the effect would not be of material consequence, would prove more unfavourable than useful ; in the
first place, because the enemy would probably pay little attention to them, and of course the object of the diversion would

The theoretical principle of the lines of operation appears to result of itself from the preceding

fail; in the second place, because the army would be exposed to the inconvenience of dividing, without that inconvenience, and the consequences which might follow being compensated by any advantage that might induce it to incur the danger. And I mean by well combined diversions, only such as, besides the qualities above mentioned, are directed against a point where the enemy cannot anticipate them without making a great movement that would put him into danger. In fine, a diversion ought to be so planned as that the enemy shall not be able, at the point on which it is made, to detach against the troops employed superior forces, which, too, after destroying these troops, might immediately return to their former station, and thus obtain a double advantage.

As diversions prove so advantageous to armies that know how to employ them properly, it is natural to conclude, that whether acting offensively or defensively, too much care cannot be taken to foresee all those which an enemy may attempt, in order to devise such means as will render them abortive. The knowledge of your own frontiers and those of the enemy, the length, the direction, &c. of your lines of operation, which must make you acquainted with the points of them most open to successful attacks; the strength of the enemy which guides to an estimation of that of the detachments he will be able to employ for making diversions; lastly, the knowledge of the country in which you act, should indicate to you the points that may be endangered, and the measures to be taken to frustrate the enemy's designs; whether by stationing at those points troops, which by their number, the advantages of the

observations on a single line of operation; and
by all that has been said I am fully warranted

ground, or the resources of art, may be able to check him;
or by making a diversion yourself, which shall force him to
retreat.

The king of Prussia states also, as one of the surest means
of discovering the enemy's projects, that of observing the
places where he fixes his depots of provisions before the
opening of the campaign. It is true, that inferences may
be drawn from this proceeding, particularly when the war
is between two powers, which, by their respective topogra-
phical situations, have necessarily fixed points for the establishment of their magazines, according to the part of the country, against which they mean to act: it would, however, be
imprudent to depend entirely on this observation; for, the
enemy may intend to mislead, by making preparations at
several points, and on a side opposite to that where he designs to act. It was thus, that Mareschal Saxe deceived the
allies in 1748. He collected all the stores for a siege at
Antwerp, made great demonstrations on that side, then went
and invested Maestricht. The manner in which the French
army was assembled at the opening of that campaign, was
truly a masterpiece, and cannot be too much studied.

I will conclude these remarks with observing, that it is particularly in countries where there are lofty mountains, that
the movements of the enemy are to be attentively watched,
and that an army is not to be imagined to be in safety, because the principal roads and openings have been secured;
for, if the enemy is well acquainted with the face of the
country, he will steal in among its posts, take positions from
which he will not be easily dislodged, intercept its provisions, and perhaps cut off its retreat, in case it should be
obliged to fall back.—COMMENTATOR.

D

in concluding this chapter with the following
positive maxim of *Strategics*,*, to wit;

"That, in a defensive war, an army must not
oppose the enemy in front, nor yet suffer patiently
its enterprises and attacks; but must choose a la-
teral position, and, avoiding the front of the enemy, become itself offensive, by harassing his rear
and flanks, and particularly by enterprises against
his convoys†. Such must be the conduct of a
defensive army, whether it aim at diverting the

* This term is explained at the beginning of Chap. X.

† In the commentary relative to the formation of plans
of campaigns, I have spoken of a *strict-defensive* to which an
army may reduce itself; and as this expression seems, in
some degree, contradictory to the principle here laid down
by the author, I think it necessary to explain myself. By
a war strictly defensive, I understand a war carried on by
an army which has no intentions to make conquests, and
to establish itself firmly in the enemy's country, which it
might attempt, according to circumstances, in an offensive-
defensive war; but merely to defend its own territory. I
am far, however, from meaning, that this war, though I call
it strictly defensive, excludes the offensive movements of
which the author speaks, and which, on the contrary, I
consider as being the best means of defence, when they are
well combined. My opinion, indeed, on this subject, is obvious in the commentary I allude to, and in entering
into a more particular explanation here, I am entirely actuated by the desire I feel to leave nothing in this work
in the slightest degree obscure.—COMMENTATOR.

enemy's attention so as to be able to attack his rear and flanks, with advantage, or at keeping him in his posts, while with as great a part of the troops as can be spared, it acts against his supplies, and even invades his country; operations which, as may be easily conceived, are all to be carried on in his rear."

In what has been already said we find this principle demonstrated; and generally, in future, I shall begin with the examination of the principle, then adduce the proofs, and conclude with establishing the rule, as the completion of the edifice.

CHAP. II.

Of Lines of Operation which, proceeding from the Extremities of a right-lined Base, form, in meeting at the Object, an Angle less than ninety Degrees ; and of those whose Base is a Segment of a Circle likewise less than 90 degrees.

THE disadvantages of such a base are almost as great as those which we have been examining. In the first place, when the two principal roads through which the convoys pass, form together, and relatively to the object of the operation, an angle less than 90 degrees, the base is insufficient.

The army D (*fig. 2.*) can make no progress against the object C, when the enemy E opposed to the line of operation B C; approaches a point of the triangle A C B: for D is obliged speedily to send a detachment back towards F, in order to cover the line of operation B C; not being able to cut off the communication of E with its magazines; for, as these are, or ought to be, in the environs of G, they are of course covered by E. Thus, as in the case of a single line of operation, the army D is compelled to act on the defensive by this movement of the enemy in his rear. Indeed, D here has the advantage that E is not so situated as to be able to cut off the line of operation A C; for, if it should attempt to advance to that line it would be in danger of being cut off itself by D, and separated from the magazines which it has in the environs of G. Besides, allowing that E should cross the triangle A C B, and reach the line of operation A C, that line, it is true, would be then cut off, but B C would be free; so that what E would gain on the one side, it would lose on the other. I must, however, observe that the enemy being in his own country, he ought likewise to have magazines at H, by means of which he might march a detachment I, that should cut off the line of operation A C, as E would cut off B C. D would then be under the necessity of marching detachments against I and E, and would consequently be exposed to difficulties similar to those I detailed in the preceding chapter, when speaking of the disadvantages resulting from a single line of operation.

Further, the angle A C B formed by the two lines of operation A C, B C, being acute, the

space between these two lines, particularly towards the angular point of the angle, is very narrow, and it would be easy there for the enemy
to cut off, at the same time, B C, A C, and
every other line of operation, such as K C,
extending from the base A B towards the object
C; consequently lines of operation of this kind
have not a sufficient base.

In fact, would not the enemy E be able there
to place himself in the rear of the army D, in such
a manner as totally to intercept the communication with its base A B, and to have himself
that base behind him? A decisive operation,
which he might effect without much danger;
for, he would draw his supplies from his flanks
G, H; and it would be easy for him by small
corps of observation, to check the garrisons in
the fortresses A, K, B, situated on the base, if
those garrisons be not strong; at all events, it is
a precaution that would secure him from ever
being entirely surrounded. Thus, should the
enemy E be completely worsted by D, and
driven from the position which he had taken in
its rear, it does not follow that he would be ruined; for, besides that he would, in some sort, be
covered by the small corps of observation which
we spoke of, from the enterprises which the for-

tresses A, K, B might be disposed to attempt against him, he would be far enough from those fortresses, to retire without danger to the environs of H and G. With respect to the army D, the best thing it could do, to rescue itself from its dangerous position, would be to make a desperate attack upon the enemy, and at the same time a movement towards his magazines at G and H. But this army would not have been in so critical a situation, in which it is exposed to be completely fenced in, if its lines of operation had been supported by a sufficient base.

Let us now suppose that the base, instead of being a right line, such as A B (*fig.* 2.), were the segment of a circle as A E B (*fig.* 3.) less than 90 degrees. The only advantage that would result from it to the army D over the right-lined base of the triangle in fig. 2, would be, that the two lines of operation A C, B C commencing at the extremities A and B of the segment, being shorter than they would be if the base were a right line, as shown by the dotted line M E N, the army would be nearer to its principal magazines, situated, as it is natural to imagine, at the extremities A and B of the segment. But then, this advantage is counteracted by a consequent disadvantage; for, the shorter

D 4

the lines A C, B C are, the nearer will be the
points A and B to each other, the angle formed
by these lines remaining the same number of
degrees: now, the closer the fortresses we suppose at the points A and B, the narrower must
be the base, and the enemy consequently will
have little more to fear from these two fortresses,
than if there had been but one, his security increasing in proportion to the abridging of the
base ; for, the shorter it is, the nearer do we approach the hypothesis of operations on a single
line. We see, then, that on the opening of the
angle C, which I shall in future call the *objective angle*, depends the degree of safety in the
operation.

It is to be observed likewise, that although the
point E, whose situation in respect to the points
A and B gives the base the form of a segment,
is perfectly protected by those points from an
attack, still the line of operation E C is not
more protected than the line K C in fig. 2.; it is
therefore exactly the same thing here, as if E
had no existence, and A C B were an acute
angled triangle, whose base would be the line
A B; and as the cause that would endanger the
line E C, is to be found in the smallness of the
objective angle C, I conclude that lines of ope-

ration, whose base is the segment of a circle less than 90 degrees, are liable to the same difficulties as those which, commencing at the extremities of a right-lined base, would form, in meeting at the object, an angle likewise less than 90 degrees.

CHAP. III.

Of Diverging Lines of Operation.

IT is a settled rule, that lines of operation
diverging from a central point to a circumference, or from a smaller to a larger arch, or from
a right line to a circumference, are indispensable,
when an army intends to occupy an undefended
country. The Tartars follow this method when
they intend to overrun a country. But the
Tartars attend little to lines of operation, of which
they have no need. They carry their provisions on their horses, which is a step beyond the
Romans, who had theirs in their camps. The
Tartars are acquainted only with lines of march.
The way they proceed, when they inundate
a country with their innumerable hordes, is
this: the general body divides itself into two,
these again into four, and so on continually
dividing as they advance, till they at length
cover the whole country. Baron de Tott tells
us, that he saw them in this manner ravaging
New Servia.

It would prove a total ignorance of the art of war, to commence an offensive campaign with these divergent operations in a defended country; for, they do not secure an army from being surrounded; its rear and flanks are always uncocovered. Its lines of operation are not secure, whereas the enemy has nothing to fear for his. The following examples, illustrated by figures, will fully demonstrate these assertions.

The line of operation C D (*fig.* 4.) is really better secured by C E and C F, that is to say, by detachments near E and F, than two lines of operation like those examined in the preceding chapter would be; I mean such as form an acute angle. But the other lines of operation C E, C F commencing at C and directed against E,F, are not protected from the enterprises of the enemy's troops stationed at A and B who will attack them in rear of the invading army, and intercept its convoys, secure of their own retreat to their posts A and B. If, in addition to this, the fortress C is not strong enough to sustain a long siege, all the advanced corps must fall back, and make their retreat, to avoid destruction, the moment the enemy, marching from A and B, menaces C*.

* Nay, even should C be able to sustain a long siege, the detachments sent towards E, D, F, would not be the less obliged to fall back to relieve it; for, the enemy who be-

It would not even be necessary, that the enemy should have detachments near A and B ; for, if he only marched towards those points on a periphery, with the troops near E and F, the detachments sent from C towards E, D, F, would be obliged to make a sudden movement in a defensive parallel towards A and B, and to form a narrower circle round the point C.

The enemy's detachments on the periphery, would always have it in their power, by a junction of several of them, to fall on one of those that had marched from the point C, and to crush it, if the distance were not too great: for, though the detachments from C acting on a narrower arch than the enemy does, might consequently effect a junction more speedily, yet a march gained by night would make matters equal. The enemy, from the extent of his base, which comprehends the arch A D B, has nothing to fear for his rear or his supplies. He can never, in this hypothesis, run a risk of being cut off. But the corps from C have not the same advantage; for, if they concentrate themselves, they expose such

sieges C, has most probably invested it; and if C is the only point whence the convoys by which the detachments are supported come, how are they to exist during the siege, since, as the town is surrounded, the convoys can no longer come out to them?—COMMENTATOR.

of their convoys as should follow the lines C E,
CF, and find themselves, in the end, reduced to
a single line of operation, C D.

The dispersion of the forces towards several
objects, prevents the acting against any with the
necessary energy. The army is weakened, and an
opportunity is given to the enemy to destroy it
partially: success cannot be expected, but by
marching on every point, forces superior to those
of the enemy. Numbers decide all, and it will
be for ever true that strength rises from union and
weakness from disunion. It is the same in this
case, as with a man who undertakes a thousand
things; none of them turn out well.

If an army, to cover its lines of operation
against the enterprises of the enemy from his
posts A, B, should assemble in detached corps,
and form, as it were, a wall of troops C D (*fig. 5.*)
opposed to the periphery A B, it would be so
weakened as to be unable to undertake any thing
of importance, and the certain issue would be,
its defeat every where: but this is so great
an absurdity, that it is useless to dwell on the
subject.

Two detachments, as D, E (*fig.* 6.), would, no
doubt, be sufficient to watch and to awe the two
fortresses A, B, and at the same time the troops

which, posted in the environs, might take in the
rear those that were directing their operations
towards the points F, G, H. But, this is falling again into the fault of too great a division
of the forces, and the army acting offensively in
the direction of F, G, H, would be too much
weakened by it. Besides, (and this is the principal consideration) it would not, if the enemy
placed at F, G, H, were to march his troops towards A, B, on the periphery A B, prevent all
the corps employed in the offensive operation
from falling back to the detachments D, E, and
thus changing the offensive into a fruitless defensive war.

If even diverging lines of operation had two
fortresses C, C (*fig.* 7.) for a base, yet, should
these places be situated so near to each other,
that the lines of operation, proceeding from those
two points, only formed with the object G an
angle less than 90 degrees, nothing would be
gained; for, the detachments sent on the offensive against G, would be liable to all the disadvantages already deduced, in the preceding chapters, from lines of operation forming together an
acute angle; and the detached corps, as F, H,
would have to bear all the inconveniences of the
divergent lines of operation.

Now, the objects towards which lines of operation of this kind are directed, ought, relatively
to one another, to be so situated, that a line
which may be supposed to cross them, may nearly
have the appearance of the arch which I have described, in the figures, by A B. In this arch terminate the lines of operation, which, like so
many rays, issue from the fortress, the principal
point of operation, as from a centre. A divergent operation cannot well have more than one
centre; but say it has two: in this case it
would be necessary that the two fortresses, whence
the lines proceed, should be close to each other.
They would then be no less exposed, as I have already demonstrated, than if they proceeded from
a single point.

Further, it is clear enough that the enemy,
setting out from A and from B (*fig.* 8.) may
make a diversion in the country of the army C,
while that army is acting offensively against the
points D, E, F; for, this mode of defence may appear to him more easy, than to place himself on
the rear of the detachments marching to D, E, F,
intercept the convoys, or undertake any thing
against C.

Indeed the attacking army, setting out from
the point C, will probably have fortresses opposed

to those of the enemy at A and B; in which
case a diversion in its country would become
more difficult: but those places may be also attacked. Finally, the enemy would likewise have
the resource, and it would be his best plan, to
project an enterprise against C, or against the
lines of operation. More; if he should be sufficiently strong, attempting all the three would
but the better enable him to defend himself successfully.

From these considerations, it is obvious, that
before an army advances towards the arch A E B,
it must make itself master of the fortresses, which,
according to my supposition, are at the extremities A, B. This would be an uncertain operation,
if the lines of the offensive army's convoys were
all to proceed from C; for, in that case, the enemy posted at D, E, F (*fig.* 9.) would be able to interrupt them, by advancing to G or to H, as A
or B should be besieged: those convoys should
therefore come from I or K, situated in the
country of the besieging army, which would certainly have, within reach of one of those two
points, a fortress containing a magazine. This being
the case, I consider the capture of A and B as
always possible: it would be attended with
greater difficulties, if K and I were not fortified;

for, while the offensive army besieged one of the fortresses, B for example, the enemy occupying A would make a diversion in the country of that army at K, and take it in the rear, if not stop‾ped by a fortress.

Once master of A and B, and having (to use a trite expression) elbow-room, the army may take A, C, B for a base, and thence project its lines of operation. It is, however, still necessary that A and B should be at a proper distance from each other, that the lines of operation, taken from those two points, may, in meeting at the object, form an angle not less than 90 degrees. If the army has sufficient strength to besiege both A and B at once, it would be right to try it; because it is an invariable maxim, that the most that can be done, without the inconvenience of weakening and dividing too much, should be undertaken at the same time. But, if those places can only be attacked one after the other, as will most frequently be the case, it will be proper, should there be no particular cause to induce a different conduct, to attack the stronger first: for this reason; that the army is always more powerful and better prepared at the beginning than at the end of a campaign.

Let us conclude with establishing, in the fol-

E

lowing terms, this principle of *Strategics :* That if a base is so long, that the two lines of operation from its extremities, form in meeting at the object of the operation, an angle of 90 degrees, an army may advance in perfect safety; but if not, that it would be imprudent.

CHAP. IV.

Of Parallel Lines of Operation.

LET me first observe, that if an army, when marching on several lines of operation parallel to one another, directs the greater part of its forces against one object and makes demonstrations against others, only to divide the enemy, this kind of operation is entirely different from what I mean by *parallel lines of operation:* but if a serious attack be made on each point, I call this parallel operations.

I have shown that a sufficient base is necessary before any offensive operation may be undertaken. This indispensable condition secured. the army may advance without danger on parallel lines of operation, so as to attack at once several objects, situated on a line that is also nearly parallel to the base. Let us see; however, if this conduct would lead completely to the end proposed.

Let us suppose the line A D (*fig.* 10.) to be a base on which the fortresses A, B, C, D, stand; and that from those four fortresses, on as many

lines of operation, the corps 1, 2, 3, 4, set out
on their march towards the objects E, F, G, H,
situated on a line parallel to this base. According to the order of things, there must be
one of these four objects E, F, G, H, the
conquest and possession of which would be
particularly desirable, and against which more
force should be employed than the subjec-
tion of the others would require; for, it is
impossible that these several objects can be
equally important, but one of the four must,
in a greater degree than the others, be the key
to the enemy's country. It is in this as in the
point chosen, before a battle, in the enemy's position; a point called the key of the position,
and on which the chief force of the attack is directed. The art of distinguishing this point has
been always greatly esteemed in generals, and it
particularly constitutes the military *coup-d'œil.*
In like manner, the art of discovering the capital object among those against which the
operations are carried on, an object which I
would fain call the *Strategic key,* is not, con
sidering its great importance, a perfection of
small moment in the leader of an army.

I shall in the first place observe, that as it is
indispensable to concentrate a greater force to-

wards the more important object than towards the others, it would naturally follow that there would not be sufficient to master all ; unless the army possessed an extraordinary superiority over the enemy. In this case (I mean, where the enemy hardly dares to show himself in the field) the best way to decide the contest at once, would, no doubt, be to advance into his country, in parallel lines of operation, with an extensive front, and to occupy and conquer as much territory as the army could spread over. But this supposition is not admissible according to rule. The art of combating becomes useless when an adversary has it not in his power to defend himself; for, its only end is to teach how to act when opposed to an able and active enemy: besides, in our days, it is only great Powers that wage war with one another, so that there can never be a very great disproportion of strength.

An attack widely extended, by means of parallel operations, would therefore be ineffectual against an enemy that could defend himself, because the offensive army would never have in time the force necessary to make an impression and reduce the places it has in view; it would also be dangerous, were the enemy to change the offensive into a defensive war, as he ought to do.

For he will always have it in his power by

concealed or forced marches * to concentre his
forces, to fall on one of the corps marked 1, 2,
3, 4, to take the others in the rear, or to act

---

* Marches that are concealed from the enemy, contribute
highly to the success of military operations; but what talents, what precautions do they not require in the Com-
mander-in-Chief, who conceives the plan, and in the officers
to whom he commits the execution of it! It is not enough,
that the former possesses a perfect knowledge of the country
through which the marches are to be made, of the character
of the enemy, of his strength, of his situation, &c. so as to
be able to determine, with certainty, the points to which
they should be directed : it is also necessary, by calculating
the distance of those points, the nature of the roads leading
to them, that of the troops to be marched, the different obstacles which they may meet in their way, &c. to ascertain,
with the greatest precision, the time required by each column to reach its appointed post; it further behoves him to
remove those obstacles as much as possible, and to provide
in the most secret manner the indispensable supplies for the
subsistence of the detachments. The difficulties increase if
the detachments begin their movements within reach, or in
presence of the enemy ; for, besides that the march must
be then masked, by leaving in front of him a body of troops
which may prevent his perceiving it in time, and which,
when the movement becomes general, is to join the army
in silence and with all the celerity possible, it will be necessary to employ an infinite number of other measures
dictated by prudence, and this the more imperiously, that
without their concurrence, the army would not only be liable
to fail in its enterprise, but would perhaps find itself in a
most dangerous position.

against their lines of operation.   It being as impracticable to penetrate between two corps, as
in a curtain between two bastions, the º outer

Then, what attention is requisite in the commanding
officers of the columns, in punctually observing the instructions given to them, in maintaining in their respective
divisions the greatest order, and, if near the enemy, the
greatest silence; in never accelerating, through a mistaken
zeal, or retarding through an unpardonable negligence, their
arrival at the point to which they are to march! for, the
success of operations of this kind depends essentially upon
their being executed with coaction and the consequences may
be equally fatal, whether a part of the troops arrive too
soon or too late: consequently, how careful should these commanding officers be not to lose their way, particularly in a
march at night, and how many precautions should ·they use!
such   for example, as arresting every suspected person ;
taking, in the enemy's country, hostages from the villages
through which they pass, &c. to remove every possibility of
the enemy's being informed of the march intended to be
concealed from him, till he may learn it without detriment
to the army,

An inexhaustible fund of instruction, relative to this interesting branch of the art of war, is to be found in Mareschal
Turenne's campaigns in Alsace.   His long march in 1674,
when he left Upper Alsace, was a masterpiece.   It was so
well concealed from the enemies, that the division which
they had at Sintzheim was defeated, before it was even suspected that the French were in motion.   The invasion of
Bohemia by the Prussians, in the beginning of the campaign
of 1757, likewise deserves the highest praise.

COMMENTATOR,

E 4

corps 1 and 4, would be attacked with the greatest advantage: and when these are repulsed, the inner ones would be compelled to put themselves on the defensive, and face to the flanks in order to cover their operations, and avoid a rout; they would likewise find themselves under the necessity of extending detachments backward to their base, for the greater security of their convoys. Would not this then conclude the offensive war on their side? Yet I will venture to say, that the offensive army has the advantage of a sufficient base, so that the enemy dares not attempt, as in the preceding cases, to turn it in a position considerably advanced, for fear of being cut off himself. It is possible also, that he may be compelled by the fortresses A, D, to retreat precipitately to his own base.

On the other hand, if the intervals between the corps 1, 2, 3, 4, were too great, the enemy would easily drive back one of the middle divisions, without the slightest danger to himself, and the adjacent lines of operation would be then exposed to his attacks before either of the wings 1 or 2 could come up to their assistance. But indeed, before this could take place, the intervals must be of such an extent, that each of the lines of operation might be considered as forming a separate operation.

It appears then, that the more an army is divided, and the more it attempts to act effectually against several objects at the same time, the more defective the organization of its base, and the less is it able to effect any thing decisive. It is rendered every where too weak to resist an enemy in any degree concentrated, and will be beaten and driven back, division after division, if the adversary knows his business.

Lastly, were the enemy to collect his forces at any point of the objective line E H, and were the offensive army also to concentre itself directly opposite to him, supposing his movement to be known in time, the consequence would be a succession of parallel marches, during which it would be impossible to undertake any thing of importance. Marching in this manner flankwise, sometimes on one side and sometimes on another, is, of all manœuvring, the least advantageous for him whose aim is to attack and conquer; and if this game lasted long, he would be reduced to a defensive war. I conclude, then, from the observations I have now made, that unless there be a decisive superiority, parallel offensive operations are not a means of success.

CHAP. V.

Of Lines of Operation which, proceeding from the
Extremities of a Base in a right Line, form, in
meeting at the Object, a right Angle, or an ob-
tuse one; and of those whose Base is the Segment
of a Circle of 90 Degrees or more.

WE have seen that a single line of operation
is insufficient to attain the end of an offensive
operation; that it is the same with several lines
which, in meeting at the object, form an angle
less than 90 degrees; that diverging lines of
operation are not more exempt from disadvantages; that it is necessary to establish a base of
a certain length, and that, in most circumstances,
it is not prudent to go forward on parallel
lines of operation; lastly, that there is always
a principal point which it is indispensable to
carry, before any other enterprise be entered
upon. We are now going to examine the case
where two lines of operation, proceeding from the
extremities of a base in a right line, form,
in meeting at the object, a right angle, or an

obtuse one; and that, where the base of those lines is the segment of a circle of 90 degrees or more. This is the most perfect plan for offensive operations. But the proposition requires to be demonstrated.

The army E, (*fig* 11.) acting from the base A D B of the right angled triangle A C B, towards the object C, has no occasion to fear being cut off, or that its convoys will be interoepted; for, though the enemy may, indeed, cut off the lines of operation B C, or A C, according to the side from which he comes, he cannot possibly cut off the line C D, or any other, either between B and D, or A and D. For, if he advance on the rear of the army E, crossing the lines of operation B C and A C which proceed from the extremities of the base A D B, he will himself be cut off from his principal posts. These, in fact, can be placed only at G, H, I. But if the enemy advances as far as the line C D, or only to the point K, the army E may very easily, by a detachment, cut off his retreat to G, and the fortress B might, at the same time, be able to intercept it towards H and I: so that he would himself fall into the snare he had laid for others.

If the angle A C B be obtuse (*fig.* 12.) the army E is but the more secure in its operations against the object **C**: the greater the objective angle formed by the two lines of operation proceeding from the extremities of the base, the longer consequently is that base, and so much the more securely may the attacking army pursue its operations against the object it has in view; for, it has nothing to apprehend from the diversions of the enemy on its rear, nor from his enterprises against its convoys *. The enemy F,

* The advantages which the author here deduces from a long base in favour of offensive are also applicable to defensive operations. I must explain myself. It often happens that an army, which at the beginning of a campaign marches into an enemy's country to act offensively against a determined object, is afterwards compelled, in consequence of some considerable check, or from other circumstances, such, for example, as a powerful diversion in his own country, to fall back, keeping itself on the defensive, and even sometimes to retreat to its own frontiers, which constitute its base. It is therefore the rule, with every prudent general, who wishes to be prepared against all that may happen, not to advance into an enemy's country, leaving behind him any fortified towns, any forts, or even entrenched posts, which, in case he should be obliged to fall back, might be in a situation to attack him, and perhaps even to cut him off: in short, a general should keep all the country behind him free; at least that which lies between his lines of operation. Now, as a long base gives him this advantage when he acts offensively, it is clear that he may avail himself of it in the retrograde movement he may be compelled to make, supposing him, from whatever reason, obliged to fall back, keeping on the defensive —COMMENTATOR.

as I have already observed, would himself be cut off from his magazines G, H, I.   A detachment from the fortress B would soon force him to return in haste to his base.   Even supposing that he dared to cross the line C B, and on the other side the line A C, the army E would still have the line of operation C D, and, in general, all the lines of operation proceeding from the base between the points of the extremities A, B, by which the army E would pursue its designs without any obstacle.   Indeed, however long the base, however obtuse the objective angle, the two lines of operation at the extremities, are always exposed to the enemy.   But even were there no other fortresses in the base than at the two points which terminate it, an army acting offensively would still have it in its power to establish on it, in perfect safety, a number of magazines whence it might draw its convoys; for this purpose it is sufficient that the base be long enough; a condition that will be fulfilled, when the dimension of the objective angle is at least 90 degrees.

Neither can the enemy make any diversion in the country of the army E; for, it is impossible for him to penetrate between D and B. He might, indeed, do it on the side of B; but then, the base of the army E projecting beyond his, he would find himself in a dangerous position, which shows how important it is to have a base of a sufficient length, defended by good fortresses *

* Another advantage arising from a long base is that of alarming the enemy in several points, and of obliging him, consequently, to divide his forces, so that sometimes he may not be able to assemble them in time to obstruct success in the first object, at the commencement of an offensive operation.

Some further illustration of the theory *of the base* will perhaps be thought useful by the reader.

We see, by what has been said, that the base determines the establishment of the magazines; and also, that the distance to which an army may go from them, without danger, is limited; for, if it goes too far, its lines of operation will become so long that an enemy, who understands manœuvring, will easily be able to cut them off: it follows, therefore, that according as an army acting offensively proceeds into the enemy's country, it is under the necessity of forming new bases in order to be within reach of its depots, and that it may not exceed the distance to which it may go from them, not only without discontinuing to cover its lines of operations, but also without depriving itself of the means of receiving in time the supplies of every kind which it may stand in need of; so that, according to the extent of country this army conquers, will it be obliged to establish base after base.   Now as in a siege the first parallel ought to be more extended than the second, and the second than the third, to embrace and cover the flanks of the besiegers as they gain ground, in like manner should the base situated in the country of the offensive army, and which I compare to

I have not determined invariably the dimension of the objective angle, because it was
the first parallel whence the besiegers proceed at the commencement of a siege, have extent enough to cover and secure by its extremities the flanks of the other bases established in the enemy's country by the offensive army as it
advances. For, if this base is but of small extent, the others,
which it must exceed, and which of course must be still
shorter, will expose the army advanced to all the inconveniences demonstrated by the author to arise from bases not
sufficiently long. Let me be permitted to observe, by the
way, that this analogy between fortification and the operations of war in general is not confined to the case before us,
and that if, in all armies, the young officers were as
convinced as they ought to be of the great affinity that exists
between fortification and the art of war generally, they
would banish the mistaken notion, so fatal to their instruction, that this science belongs exclusively to the Engineers
and Artillery, and would be more anxious than they usually
are to learn, at least, the leading principles of it. I leave it
to intelligent officers to determine how far my observation
is just, and shall resume my illustration of the theory *of the
base*.

I have said that an army is under the necessity of forming new bases, according as it proceeds into the enemy's
country, and I have given the reasons for it. If it be
master of its adversary's frontiers, and those frontiers be
provided with fortresses, which by their nature and position
are proper for establishing a base, it is, no doubt, natural to
convert them to that use: but I must observe that all fortresses are not equally useful in themselves to the army
which has taken them, and that there are some which it
ought rather to demolish than to keep when it his

enough for me to lay down the principle, which
may serve as a rule in all cases.   In this respect

taken them,   If, for example, a fortress, from which
the enemy derives great advantages while he occupies
it, is not indispensably wanted to secure your magazines,
and your establishments in general; to cover your lines of
operation, your communications with your country, to keep
the enemy at a distance from it, and to facilitate your progress in his own, demolish it ; especially if there be any
fear of his retaking it.   Even sacrifice, in this case, the temporary advantages you might reap by this fortress, if they
are of less importance to you than those which the enemy
would enjoy in retaking it.   If a fortress, of which you
have made yourself master, because it obstructed your operations, obliges you afterwards to weaken your army for the
garrison it requires, without giving you advantages that
overbalance this inconvenience, demolish it likewise.   Lastly, if you have made yourself master of a fortress that is the
key to a country, which political considerations will prevent
you from keeping at a peace, and with which you may, at a
future period, be again at war, take advantage of the opportunity, and demolish it also ; unless your subsequent operations make it absolutely necessary for you to spare it.
Should I be charged with carrying this foresight to excess,
I would reply that in war we are bound, while we attend to
the duties of humanity, to do the enemy all the necessary
mischief we can ; that the duties of humanity relate only
to the preservation of the life and property of individuals;
and that, in short, the safety of the State we defend is infinitely of more consequence than the preservation of the
symmetrical order in which the stones that form the ramparts of a fortress are laid.

circumstances necessarily modify the principle, and I have traced the bounds between bad and good within a few degrees; more precision would have been pedantry. Thus, when the objective angle is 90 degrees, the base of an operation is essentially good, though I have laid it down as decidedly so, only when the angle exceeds that number of degrees. It is only according to circumstances, then, that the number of degrees which the objective angle ought to have can be determined *   Perhaps I shall be told that, ac-

If the country conquered have no fortress, the first thing to be done is to consider of the means of entrenching the towns, villages, and posts, which may best secure the magazines and lines of operation, as well as the communications of the army acting offensively ; and not to determine on the points to be put in a state of defence, till after maturely reflecting not only on the actual situation of things, but also on what the enemy may do, if circumstances enabled him in the sequel to assume the offensive. For, besides, that these points constitute the base from which the lines of operation will proceed, whether the design be to penetrate further into the enemy's country, or it should be necessary to oppose his offensive operations, in case he should have received reinforcements, they may also become the line of the winter-quarters, which it is essential to secure. —COMMENTATOR.

* To recapitulate in a few words, and in a clear manner, the principles demonstrated by the author in this chapter and in the second, we may say ; that an offensive operation

cording to my doctrine, it would require nothing more than the use of an astrolabe to discover
if it be prudent to commence a siege, or if an
army is sure of not being separated from its convoys in a given position: I answer, that it is most
certainly necessary to measure the objective
angle on a map, and to weigh every thing maturely, previous to engaging in an offensive operation; but as to jests, I consider them as un-
worthy of my attention.

Let us now suppose that the base is the segment of a circle of 90 degrees, such as A D B
(*fig* 13.): the only advantage which a base of
this kind would have over a right-lined base
(the objective angle being likewise 90 degrees)
would be, that the part situated between the circular base and the dotted line A B being completely out of the enemy's reach, the lines of

is in general supported by a sufficient base, only when the
length of the base is such that the angle formed by the
meeting of the lines of operation, drawn from its extremities
to the object, is at least 90 degrees; that if that angle exceed
90 degrees, the base is the better; and the more, on the contrary, it decreases from 90 degrees, the more will the base be
insufficient in most cases: but that, as it is impossible to fix
the degree of extension the objective angle ought to have,
since the principle must necessarily be modified by circumstances, it is requisite, in calculating that extension, to be
guided by circumstances.—COMMENTATOR.

operation issuing from between the two points of the extremities A, B of the circular base, such as D C for instance, would be perfectly secure. On the other hand, it is disadvantageous that those lines are longer, by the whole space lying between the dotted line A B and the arch A D B, than they would be if the line A B were itself the base; for, the longer the line of operation, the more tardy, irregular, uncertain, and expensive will be the conveyance of the supplies. It would be better then, that the dotted line A B were the base, and that D were nearer the object of the operations of the army E. For the same reason, it would be disadvantageous to take the line F D G for a base, as it lies at so great a distance from the object C.

If the segment which forms the base exceed 90 degrees, and the objective angle be very obtuse (*fig.* 14.) all the lines of operation will be completely protected. But so would they be also, if the base were straight (*fig.* 15.) because the security of offensive operations depends solely on the extension of the objective angle. In fine, if the base form an arch projecting towards the enemy, such as the arch A B (*fig.* 16.) it is clear that only the part *d, e,* of that arch, can be truly considered as the base: for A and *d* on one side, B and *e* on the other, being on the same line of operation, it is natural that the army acting against C should draw its supplies from the magazines *d* and *e*, which are the nearest to it. By this we see that, although the security of an offensive operation depends particularly on the extension of the objective angle, still the direction of the base is not a matter of indifference.

I shall conclude this chapter with laying down the following principle, which is the result of the preceding discussions; to wit: " that previous to acting offensively against a determined object, it is necessary to have a base of such an extent, that the angle formed at the object by the meeting of the two lines of operation which proceed from the extremities of the base, may be at least a right angle, whether the base be a right line, or the segment of a circle."

CHAP. VI.

*Of the most advantageous Form and Direction of
a Base.*

HAVING considered the length of the base,
we now come to treat of its direction. This
subject has never yet been properly discussed,
and from its importance it deserves great attention; especially as, having already spoken of
offensive operations beyond the base, and pur-
posing to speak of retreats within the base, even
in the country of the attacking army, the order
of the subject requires that we should take under
consideration all that relates to the base itself,
which lies between those opposite positions.

It is clear that an arch enclosing the enemy
is the most advantageous form a base can have;
for the enemy can take no tenable position
within the arch A B (*fig.* 17.) he is taken as in
a sack, and all he can do cannot defend the
country enclosed in that arch.   All the for-

tresses within it must fall into the hands of the army that advances supported by such a base; for there is no objective angle at C, that point being in the diameter of the circle A B, and consequently all the fortresses in that circle, such as D, E, are cut off, and must necessarily fall, if the proper steps are taken. The enemy has no resource left but to make himself master of the two fortresses A, B, at the extremities of the base, in order to shorten the arch, which forms that base, and so far to free the country there enclosed. Every other enterprise formed in the interior of the arch would be abortive; and a country so invested may be considered as conquered.

On the contrary, an arch projecting towards the enemy is the worst form it is possible to give to a base; for the fortresses C and D (*fig.* 18.) are very much exposed, and the moment they are cut off, by means of two columns he detaches from E and F, the hope of saving them must be given up. There would be no means of stopping him, but by acting sideways, and advancing a little from A and B, to harass the detachments E and F on their flanks and rear; to oblige them to fall back; and to try, during that time, to take a fortress that would of course be standing either at the point *g* or the point *h*.

Thus, the more eccentric the arch, and the nearer it approaches the form of an ellipse, as A B (*fig.* 19.) the less easy it is to preserve the places situated on the part of that arch which bulges towards the enemy, such, for example, as C; the more, consequently, are the operations exposed to all the disadvantages we have mentioned.

If we now suppose a base established on two rows of fortresses in right lines, such as A, B, G, and c, d (*fig.* 20) and those fortresses so disposed that lines drawn from one to another form, successively, salient and re-entering angles, as A c B, c B d, B d G, it is clear the enemy cannot advance within the re-entering angle c B d; for, he would then be surrounded. The fortresses c, d, situated at the angular point of the salient angles, are still less exposed to the attacks of the enemy, than a place situated like C in figure 19 would be; for if d, for example, were attacked, it would be easy, from the point c, to make a diversion into the enemy's country E F: such a base, besides, possesses the many advantages to be expected from a straight base.

It is advantageous that the points A, B, (*fig.* 21.) which terminate the second line of the fortresses, should extend beyond the extremities c, d of the first: particularly, if the place most advanced be bounded by the sea, or a great river, so as not to be exposed to be turned; for then, the two fortresses A and B form a flank of the most formidable nature, beyond which the enemy could not venture with impunity : because the defensive army, protected by those fortresses, would very easily make a powerful diversion on his rear, and even in his country; particularly by concentring its forces near A and B, and afterwards falling upon the flanks of the offensive army.

But if the fortresses c, d, are not supported, but are, as it were, thrown to the winds, it is clear they stand in the way to be the first victims of the enemy, who may very easily cut them off.

I shall now consider a base, the fortresses of which are all on a line that hardly deviates from a right line. In this case, the base that extends wider than the extremities of that opposed to it is the better one: thus; c, d, (*fig.* 22.) is a better base than the enemy's base A B; for it is clear that his country e, f, lies open to the enterprises that may be projected against it. Hence this rule; that, wherever it is possible, a fortress should be built fronting each of the enemy's: and, of course, it is absolutely necessary to have fortresses opposite to those which form the extremity of a base; for we have observed that it was very difficult to penetrate between two fortresses, at least if they do not stand at a very great distance from each other: thus, *g* is sufficient to cover the space between *c* and *d*, against the three fortresses A, *h*, B.

When a base *c d* (*fig. 23.*) stands obliquely to that opposed to it, A B, so as to form an angle with it, it is evident that the whole country between A and *c* lies open to the enemy; he may detach parties to ravage it, or may even occupy it with his army. This would not be the case, if the base were parallel to that of the enemy, as appears by the dotted line *e, d*; for then the whole would be covered. Indeed, *c* stands in some manner out of the enemy's reach, owing to the oblique direction of the base; because the lines of operation, issuing from A, or from the fortresses next to it, and proceeding either against *c,* or against the parts of the base nearest to *c,* are so long, that it would be very easy to cut them off. It follows then that, without detriment, the point *c* may be left uncovered, that the troops may be concentred at *d,* and march principally against B; but the enemy preserves the advantage of ravaging, unobstructed, the country between A and c *. We shall, conse-

* If the oblique direction of the base C D, to that of the enemy A B, should give no other advantage to the latter, than that of having it in his power to make incursions into the country between A and c, where certainly no magazines, no depots were established, nor even any provisions or supplies of any kind left, such an advantage might not overbalance those which c d would derive from his being able to concentrate greater forces at d, as the author says, and to act more effectually against B, which, I take for granted, is an important point. In fact, should a diversion of this kind lead to nothing more than plundering, and levying of a few contributions in an unprovided country, of no large extent, which the enemy of his own accord has left defenceless, and which is no military point necessary for him to preserve, it would be only an operation of the *petite guerre*, which can present no motive to induce an army, acting offensively, to abandon a situation in which it has very great advantages. Besides, if A B send parties into the undefended country between A and c, d might send others to oppose them, and protect in some degree that country, without depriving himself of the means of pursuing with advantage his operation against B, which is his principal object. But, if A B can march an army into the country between A and c, and then change his defensive into offensive operations, there is no doubt that c d, who could then only act with a part of his forces against B, being obliged to oppose a considerable portion of them to A B, would lose the advantages which he might have gained by his concentration towards

quently, lay it down as a principle, that a base, parallel to that of the enemy, is preferable to one in an oblique direction to it.

B, and that he would no longer be able to pursue, with the same prospect of success, his offensive operation against that point. It is in this point of view that I consider, and admit with the author, the imperfection of a base standing obliquely to that of the enemy. However, the object I have particularly had in view in this note, is, to point out again the inutility of diversions that are not directed against an object of importance to the enemy, and to show that they ought not to be considered as eligible measures to be opposed to the movements of an army acting offensively.

COMMENTATOR.

CHAP. VII.

Of Retreats within the Base: of Retreats on a Single Line, and of those which are concentred in the Form of an acute or of an obtuse Angle.

HAVING treated of offensive operations, supported by a base, and of that base itself, the nature of the subject leads us next to consider the operations of retreating within the base.

A retreat on a single line is a capital fault; for it is clear, that if the army C (*fig.* 24.) retreats from A towards B, on the line A B, the enemy, while he employs a very small part of his troops to harass the rear guard of C, and stop it in its march, may move likewise two detachments *d, d* on its flanks, that shall cut it off from the point B, or arrive there before it, in which case it would be surrounded. And more, the whole country situated to the right and left of the line A B, would immediately fall into the hands of the enemy; whereas, it is a rule, in a retreat, to cover as much country as possible.

A concentrated retreat, in which an army falls
back from a spacious towards a narrow position,
in such a manner that the two lines of operation
from the extremities A, B (*fig.* 25.) meet at the object of retreat C, forming an acute angle, or (as
in *fig.* 26.) an obtuse one; such a retreat, I say,
would have no better issue. It would be attended
with the same disadvantages which we have been
exposing in retreats on a single line. There is
one circumstance which might induce a general
to retreat in this manner; and that is, the covering an important place, a capital town for example,
by taking an advantageous position, as that at C
in the figures mentioned. That important place
would probably be situated at D. But this step
would, nevertheless, be ineffectual, if the enemy
were versed in the art of war, and in consequence acted on the flanks of the army he pursued. The best manner of covering a country
lying behind, as will be seen in the following
chapter, is to fall upon the flanks of the enemy
advancing, and by this bold movement to change
the defence into an attack. At all events, the
most absurd conduct would be to wait for him in
a defensive position, and to bear his insults.

CHAP. VIII.

Of Parallel and Eccentric Retreats.

A RETREAT in parallel lines, as from the base
A B (*fig.* 27.) in four bodies, 1, 2, 3, 4, on the
lines A C, E G, F H, B D, is better, no doubt,
than the concentrated retreats we have been examining.   In the first place, the country is better
covered by means of those parallel lines; and in
the next, the enemy cánnot so easily insult your
flanks, as you are prepared yourself to insult his,
and by that means to arrest his progress.   Besides, he is afraid of advancing too hastily, the
moment his attention is divided by the consideration of what enterprise may be undertaken
against himself.   There is still a better mode,
however; and that is, advancing a step nearer
perfection, to retreat eccentrically.

Parallel retreats are supported on the opinion
that a country is better covered when an army
leaves it straight behind, and that the progress of
the enemy is better arrested when opposed in a
direct manner.   In fact, this seems to be the
case.   But the senses often lead us into error.

They are Will-o'-the-Wisps that allure us into swamps; the present instance is a proof of it. This opinion did not stand on good ground, even among the ancients, and it rests upon much worse among the moderns. In our days we do not stop an enemy by opposing him in his strongest part, in front; but, on the contrary, by attacking his flanks, which are his weakest parts; by alarming his rear; by menacing his provisions, and his communication with the sources of his power.

Hence it follows that eccentric retreats are the best. An army (*fig.* 28.) that retreats from *a*, *b*, *c*, *d*, *e*, towards *f*, *g*, *h*, *i*, *k*, runs no risk of seeing the enemy advance into the arch *f k*, for by such a movement he would be in danger of being surrounded. This is so clear that it would be a waste of time to dwell longer on the subject.

It has long been laid down as a rule, that in retreats it is essential to divide into different columns, in order to distract the attention of the enemy: but eccentric retreats have never, to my knowledge, been established as a principle; though I think I have demonstrated that there is not an axiom of war more important.

I have already shown that by thus drawing the attention of the enemy to several points, he is kept in alarm for his flanks and rear. Besides, it naturally follows from what has been said of the inutility of diverging offensive operations, as well as of those in a single line, or in an acute angle, that eccentric retreats are to be preferred to all others. As concentric operations are most advantageous in attacking, eccentric ones must, of course, be so in defence: every thing must be contrary in two modes of war so opposite in their nature and their interests.

CHAP. IX.

Of the Result of all the foregoing Inquiries, relative to the Spirit of the modern System of War.

FROM what has been said in the foregoing chapters, it is evident, that it is more conformable to the genius of war, and the latest mode of carrying it on, that a general should make his own magazines, and the safety of his lines of convoy, the principal object of his operations, rather than the army of the enemy itself. The reason of this is, that modern armies have not the sources of their preservation in themselves, but, quite the contrary, out of themselves. In this particular, they resemble the men of our age, who place their happiness, and as it were their whole existence, in external things, and never seek it in themselves. The magazines are the heart, which cannot be hurt without annihilating that assemblage of men we call an army. The lines of convoy are the muscles of the military body, which would become paralytic, if they were cut off. But, as convoys come only on the sides and from behind, it follows, that the great object

G

of the operations, whether in offensive or defensive war, is to keep the rear and flanks of an army inviolate. Another consequence of these principles is, that fighting is to be avoided, and particularly in front. In offensive war, making movements around the enemy, and alarming him for his supplies, is a much surer way to compel him to fall back, than expelling him by main force out of his position ; for he would soon find another, where he would be firm again.

As for defensive wars, it will be very easy to discover in them how ineffectual all positions and all parallel marches are, to form a bulwark against the enemy. There is no position, however well it may be protected against an attack in front, however well chosen it may appear to cover the country you have to guard, from which you may not be very quickly expelled by the manœuvres of the enemy on your flanks, particularly if he be superior in forces *   From these

* There is no doubt, if the enemy be active and understands manœuvring, and particularly if he be superior in forces, that he will find the means of turning your position, either near or at a distance, and to take it not only in flank, but likewise in rear : now, there are very few positions which, presenting a double front, give the army that occupies the position, the means of fighting with equal advantage on either front; and even were such a position to be found an army posted there could not depend on the

truths it is that I boldly draw the rule, which is entirely new, never to wage a defensive war,

arrival of its convoys, which, on the contrary, it would be almost impossible for it to receive; particularly, if the two flanks could be turned at once. There are, likewise, very few positions which furnish the means of preventing, by shorter movements, all those which the enemy might make on their flanks and rear. Even supposing, that in a position of this nature, the commander of the defensive army, having foreseen all the offensive movements to which he may be exposed, should have previously determined upon the measures to be taken to oppose them when needful, cannot the enemy, even if obliged to carry provisions for several days, and thus to do without magazines and depots, conceal some of those movements from him, and anticipate him at the points which he intended to occupy according to such or such circumstances? In such an hypothesis, what shall the army acting defensively do? Shall it maintain itself in its position till the enemy be able to attack it? If it take this resolution, what will become of it, if the enemy succeed in his attack, as most probably will be the case? Shall this army, on the contrary, quit its position previously? If so, it will lose all the advantages of the ground on which it had counted, and may be forced to a battle where the enemy shall think proper to give it. To this add the embarrassment arising from its baggage, its provisions, the fear of being cut off from its magazines, &c. There is therefore no position, however strong it may be against an attack in front, that can be considered as a sure point of support: and if this general rule admit of any exceptions, these can only arise in the very rare case in which a position, contiguous to the object to be covered, should not

properly so called, but to change it speedily into offensive, by the simple act of falling upon the flanks of the enemy, and attacking his rear.

leave the enemy any means of manœuvring either upon its rear or flanks; such, for example, would be that which an army might take facing a single defile it had to defend; and such too would be that which it might take in front, or close within reach of a fortress which the enemy must necessarily besiege, having at the same time its flanks protected either by natural obstacles, or by other fortresses, which would not suffer them to be turned. In these cases, no doubt, there would remain no other part for the enemy to take, than that of attacking the army in front at every hazard; or, if he were greatly superior in number, of making a diversion upon some other very important point, in order to oblige the defensive army to quit its position, or to weaken itself there in such a manner that it may be attacked in front with less danger and a greater prospect of success.

Independently of the observations I have made as to the case in which a position would be turned, and as to the possibility of almost always practising this manœuvre, if we consider the chances which accident or other circumstances may throw into the hands of the enemy to make himself master of it by an attack in front, we shall be thoroughly convinced, that in a defensive war it would generally be very imprudent to depend upon a position, however strong it might be in front, to cover a country which it was of great consequence to preserve; and that nothing more should be expected from it than a temporary protection, which, no doubt, may, according to circumstances, obtain very great, although temporary advantages, for the army acting defensively.—COMMENTATOR.

Even, were the army weak, it would still be in the power of an able general, to compel a superior army to retreat and act on the defensive, by attacking its magazines and lines of convoy; and the better, that it would be enough to approach these lines, to annihilate them; that is to say, to render them useless. I shall lay it down, then, as a general rule: that an army should not take its position directly in front of the enemy; but aside of him. This maxim I mean to confirm, by devoting the remainder of the first part of this work, to the application of the rules of *Strategics* which I have hitherto dwelt upon, to *Tactics.*

CHAP. X.

The Difference between Strategics and Tactics.

BEFORE we apply the rules of *Strategics* to *Tactics*, it is necessary to define what is meant by these terms, and to mark the difference between them.

The former, called by the French *La Straté-gique*, has been almost always defined the science of the *stratagems of war*. I do not mean to dwell upon the accuracy of this definition, but I think it too limited, if nothing more be meant by *stra-tagems of war*, than what has been hitherto understood by that expression. Some, tracing the term to its origin, have denominated it, *the General's art :* this, on the other hand, is too extensive ; for, *the General's art* comprehends the whole art of war, which consists of Strategics and Tactics, sciences essentially different. These definitions, therefore, appear to me equally inaccurate.

I define *Strategics*, the science of the movements in war of two armies, out of the visual cir-

cle of each other; or, if better liked, out of cannon-reach.

Tactics are the science of the movements made
within sight of the enemy, and within reach of
his artillery.

The reader may, if he pleases, prefer the reach
of cannon to that of sight, for the bounds within
which, the movements of war cease to be Straté-
gics and become Tactics. But were I to decide
for one of them, I should adopt the reach of
sight; for the following reason: that deploying
columns in order of battle, is an operation of
Tactics; yet, it is generally done out of cannon-
reach. At Rosbach, to be sure, it was not so,
but what was the consequence?

To place the deployment of columns in order
of battle, among the operations of Stategics and
not of Tactics, would be contrary to the usual
mode of expression: it is true, that this mode cannot be said to be the only good one, considering
the unsettled state in which the characteristics
of these two sciences yet remain. It seems, however, to be logically necessary to include in Tactics, all those movements which, on account of
the proximity of the enemy, and the apprehension
it excites of being attacked while manœuvring,
cannot be made but in a position of defence,
with close ranks, and as a corps regularly formed.

This, in fact, is the best way of fixing the bounds
of Tactics.   The marching forward in order of
battle previous to an action, must consequently
be likewise considered as a part of Tactics;
yet, during the whole time it takes up, the army
is not within reach of the cannon, though always within sight, of the enemy; for it would
be absurd at a greater distance, and with nothing
to apprehend, to move and advance in order of
battle.   Such conduct, besides, would prove nothing, for it would be contrary to rule.

If against the principle, that every evolution
performed by an army within sight of the enemy
is a part of Tactics, it be objected; that it would
be possible for him to have it observed by patrols,
at the distance of two leagues, or to see it at
that distance, either in a very spacious plain, or
from a very elevated spot; my answer is, that
what is here meant, is only the possibility of
being seen from the front of the enemy's line,
and without any peculiarity in the situations.
For the most part, this reach of sight extends
further than that of cannon.

There are likewise cases, no doubt, in which
an army is so near the enemy, that, though
there is no possibility of being perceived by
him, it is nevertheless necessary to be in a
state of defence, and prepared to manœuvre;

for instance, at night, or in a wood. But, these cases are only evident exceptions to the general rule; for though the army cannot be seen, it is within the presumed distance of sight, which is what I mean.

I might still render the definition more general by saying, that Tactics are the art of placing and moving the troops, when an army is so near the enemy, that it is under the necessity of taking measures of defence against a sudden attack; that is to say, to place the troops under arms, form them in order of battle, and have them ready to fire.

But even this definition is something more precise still, when the distance from the enemy, is limited to the extent of sight.

If it be said, that all that immediately relates to a combat should be called Tactics; and if, in support of this proposition, be adduced what I asserted, in the first chapter, relative to the aim and object of military operations; that is, that all operations of which the enemy was the object, were operations of Tactics; and that those of which he was merely the aim and not the direct object, made a part of Strategics; it will then be necessary to answer the following objection: The marching in columns, as a movement preparatory to a battle, is assuredly not Tactics, but **Strategics**; for, it absolutely differs in nothing from the other operations of the latter kind: yet the enemy is its object, and that as immediately as possible. All the movements, then, which relate immediately to a combat, cannot be called Tactics : on the other hand, all those that take place during an action, and foremost the march in order of battle which precedes, are assuredly such. In certain cases, however, an action may not be the object of this march in order of battle. For, a general may manœuvre tactically before an army, and in sight of it, to make a show of attacking it, without having the least intention of it. Here we have Tactics, and no battle. How are we to get rid of this dilemma?

I shall resolve this difficulty by saying, that when troops within the reach of sight, act or *seem preparing to act* one against the other, their operations are those of Tactics; and that moving about, or marches undertaken solely for the purpose of passing from one place to another, are operations of Strategics. The former require the proximity of the enemy, and are more particularly characterized, when the manœuvring army is, or is presumed to be, within his reach of sight; the latter take place at a distance from him. The entering a camp, the passing a defile, when executed with all the forms of Tactics, proves what has been said above; for, in these circumstances, it is admitted to be possible that the troops may find themselves attacked by the enemy, not being sure of the distance at which he is. It is presumed that he is within the reach of sight, and may in consequence fall suddenly upon them; but, if he is ascertained to be at a distance, such precautions are neglected, unless it be to exercise the troops.

Encamping is not to be considered as belonging to Tactics; for, though the army is then displayed, and consequently sooner ready to defend itself than when marching in columns, this disposition is only for the purpose of being able the more easily to convert the state of rest in a camp, to a state of defence, in case of a sudden attack. While in their tents, the troops are not in a situation to defend themselves: they must, therefore, begin by breaking up the camp, taking a posture of defence, and preparing in reality to follow these demonstrations with action; which is then really an operation of Tactics. But it follows, that encamping belongs no more to Tactics, than the march which precedes the deploying into order of battle, makes a part of the order of battle itself.

Thus, Strategics relate to the positions and movements of the troops, at such a distance from the enemy, that no attack is to be apprehended, and that it is not necessary to be ready to fight; in a word, at a distance beyond the reach of sight.

Strategics include two principal parts; marching and encamping. Tactics comprise two likewise, the forming of the order of battle, and battles, or actual attack and defence. The whole together constitutes the art of war.

Tactics are the completion of Strategics; they accomplish what the others prepare: they are the *ultimatum* of Strategics, these ending, and in a manner flowing into those. The commencement of Tactics is, as I have said, the forming of the order of battle, by deploying the columns before an engagement; and, if an army is attacked in its camp, Tactics begin with the order of battle, forming in front of the camp: previous to this, all is Strategics.

The attack and defence of fortified towns, certainly make a part of Tactics; they are the

Tactics of sieges. It is true, that during the siege, the army is not always in an active, manœuvring state; but still it is much more in a defensive attitude, than in Strategic operations.

It remains for me to inquire, in this part of my work, whether the principles of Strategics explained, and I conceive demonstrated, in the foregoing pages, are not also applicable to Tactics. If they are, it will be the more clearly demonstrated, that they are the only true ones *.

* The reader will no doubt find this chapter as long as I have found it: the author might have abridged it considerably, without lessening the merit of his work.

COMMENTATOR,

CHAP. XI.

Of the Order of Battle, considered as the Base of the Lines of March, and of the Lines of Fire; and of the latter, considered as Lines of Operation in Tactics.

THE operations of Strategics begin with the establishment of the base, which may be termed also the deploying of the materials of war; and those of Tactics with the deploying in order of battle. Tactics regulate the lines of march, and the lines of fire; Strategics establish the lines of operation: so far, Strategics and Tactics are analogous. But the base is permanent, while the order of battle is never formed but for an action: and besides, the base is almost always established before the commencement of war: here they differ. Again, where there is no base secured by fortresses at the beginning of a war, it has been shown that the first operation to be undertaken, is to acquire one; so, before every engagement, it is necessary to deploy in order of battle: here we have another analogy between Strategics and Tactics.

These premises being established, I observe, in the first place, that there are as many lines of march issuing from the order of battle, which is their base, as there are soldiers in the first rank. In approaching the enemy, from these lines of march are formed lines of fire, at least in the infantry: for, since the introduction of fire-arms, in the engagements of the infantry it is not generally by the sword, or by bodily conflicts, that an enemy is driven from his position, but by balls shot to a distance by the means of gunpowder.   Thus, as Strategics finish in Tactics, so lines of march become eventually lines of compulsion, which are, precisely, what I have just called lines of fire, in the infantry.

It will be easy to apply to those lines of fire what we have demonstrated relative to Strategic operations; namely, that those which are concentric are good, and consequently, that those which are eccentric, are the reverse.   I do not, however, pretend to say, that the lines of march and the lines of fire, are analogous with the lines of convoy; and for this reason: every platoon, in advancing to the enemy, describes first lines of march, and then, lines of fire: and in

Tactics, the first operation is to form in order of battle, before any thing is undertaken; whereas the convoys, which can proceed only by frequented roads, (for it would be impossible, otherwise, to send waggons and beasts of burden to a distant army) necessarily march in columns, without the possibility of forming in line; it is therefore simply an operation of Strategics, and has no resemblance to the manœuvres of Tactics.

,,But there are cases in Tactics, that may be compared to Strategic operations, in which the objective angle is under 90 degrees, as well as to those carried on in parallel, eccentric, concentric, and divergent lines. All attacks, for example, made by a small front on a large one, are operations of Tactics analogous to those of Strategics, which are directed towards the angular point of an acute angle; and if on both sides the enemy's line extends beyond the line of the army, like C D (*fig.* 29.) the latter is thrown into dilemmas similar to those which are the consequence of Strategic operations undertaken without a sufficient base. It must of course be surrounded; that is to say, attacked in rear and on the flanks.

Why is this situation considered as so great a disaster? I think the cause of it has never been

clearly and fully explained. I have already
stated the reasons why those Strategic operations,
where the enemy is able to act on your flanks
and rear, are extremely disadvantageous; and
before we proceed further, I will point out why,
in similar cases, battles or operations of Tactics
are exposed to similar consequences. The fact,
no doubt, is universally admitted, but the original cause of it has not been investigated.

Man is so formed that he is not able to defend himself behind, or sideways. Consequently,
armies, whose first concern is to be able to defend themselves, in whatever posture they may
be placed, cannot be expected to act easily in
those directions. It follows, that by taking
the enemy in one of them, we easily conquer
him. It is a principle that a whole possesses
the qualities of the parts which constitute it;
and, of course, a battalion, or an army composed of men, is subject to the disadvantages
as well as possessed of the advantages resulting
from the form of man.

The art of war and the art of fencing are alike
in this, that it is important to draw your antagonist's attention to one part, while, collecting
your strength, you attack him in another, where
he is open. This, in fencing, is called a feint;
in war, a demonstration. They both proceed

from the same principle, and are founded on the
personal formation of man.

It is impossible to defend oneself, when attacked before, on the sides, and behind, by several
persons at once; one must yield or be beaten;
and most men, so threatened, would fly without
a battle. It is the same where the action com-
mences at some distance from the person, as in
the engagements of infantry in modern times;
for balls, which have superseded pikes, are
equally dangerous.

It is even easier in a fray, to beat off a very
superior number; for skill, supplying the deficiency of strength, may put things upon a par.
By a well-directed blow, you may knock down
an antagonist attacking you sideways, and so
quickly, that you may disengage your sides,
before others can come on to overpower you.
Skill, strength, and agility, are sufficient in
this kind of combat, to render a superiority of
number useless. Personal qualities are here
much more essential than the number of the
combatants; and for this reason, among the
ancients, the superiority of Tactics and of discipline had a different effect than with us.
It was to such advantages that the Romans
were indebted for their victories. But now,
that infantry do nothing but fire, and that

fire-arms decide every thing, moral and physical qualities are of no weight; for a child, by drawing a trigger, may bring a giant to the ground. , At present then, number, when disposed in such a manner as to be able to fire, must ensure victory Thus, when the enemy is surrounded, when each of the men who compose his army becomes a point of aim to three lines of fire, which come upon him all at the same time, he must yield or fly.

«The» number of concentric lines of fire is, therefore, decisive; and as it is not convenient to fire when the troops are behind one another, it follows that they must be deployed, in order to overwhelm the enemy with a superior fire. A greater number of men formed in column will be overpowered by a smaller number properly deployed*. We see, then, that the order of battle in line is not so monstrous a thing as the Chevalier de Folard pretends. It is rather a necessary consequence of the system of fire-arms; and as this system improved, and the principles of it became better understood, the order of battle became slenderer and

* The author, no doubt, supposes here an action supported only by fire-arms, and does not take under consideration the shock, by means of which a column might be able to overthrow troops deployed.—COMMENTATOR.

longer. The mind of man fell naturally into this conclusion. Indeed, this slender formation has generated an irremediable defect, the flanks having become excessively weak. This requires no proof. But the very defect proves the advantage of concentric lines of fire; for, it follows that the enemy is lost, if you can attack him on such feeble flanks. In fact, however weak an army may be, compared to the enemy's number, it will always have more troops than he can oppose in his flanks, and it may always worst him by means of concentric fires directed towards those points.

To prove it, let us suppose that the flank B of the line A B (*fig.* 31) is attacked by the line *c d*; it is clear that A B cannot form in a parallel direction to *c d*, if this line pushes constantly on towards A: for, A B is hemmed in, as we see, and likewise forced down in such a manner, that the whole must rush towards A. The troops that should attempt forming on the line *e f*, would not have time. Taken in front and flank by the fire of their assailants, they could never resist an attack of the kind.

Cavalry encounter similar disadvantages in similar cases. A horseman attacked on the right, on the left, and in front, cannot defend himself. The speed of horses, indeed, enables cavalry to deploy more rapidly than infantry; but, for the same reason, the enemy's cavalry acting in flank, will push sooner from the point B towards the opposite wing A, than a body of infantry can. It will, consequently, be quite as difficult for cavalry to form on the line ef; the whole will be pressed and driven, in great disorder, towards A.

The truth of these observations is so evident, and the soldiers are so convinced of it, though confusedly, that, whether infantry or cavalry, all fly when vigorously attacked in flank by an enemy. It follows, that a general should make every effort to turn the army opposed to him; that is, to bring his front to bear upon its flanks.

To the advantages in Tactics, which may be hoped from an attack in flank, are attached some in Strategics; and consequently, when the enemy finds himself out-flanked by the opposed army in a battle, he must be alarmed in a double point of view: because, if he be worsted in flank, he necessarily leaves some of his lines of operation insecure. The conqueror will have it in his power to seize his convoys, cut off the communication with his magazines, or make a diversion in his country.

It requires no great effort of imagination, to conceive all this clearly.

The lines of march and the concentric lines of fire, gained by means of skilful manœuvres in Tactics, are therefore still more important, as to, consequences, than the concentric operations of Strategics. It is owing to the advantages obtained by the concentric lines of fire, that besieged fortresses. are reduced. The fire of these is eccentric, and consequently of little effect: the fire of the besiegers who surround a place is, on the contrary, concentric, because they form a circumference at a distance from the central point to which that fire is directed; by that alone the effect is so much the more powerful, and it must in the end prevail. It is on this account too, that the sorties of a garrison rarely succeed; for, they are eccentric operations of Tactics.

According to the observations that have just been made, we see, that the best thing an army can do, and particularly when it is weaker than the opposed forces, is to attack the flanks of the enemy, unless it means to avoid fighting altogether, and to confine itself to manœuvring against his supplies, which are almost always in its power. But it is a rule, to keep a portion of your troops

in front of your adversary, when you seriously
attack his flanks; for if his line be long, he
will have time to move all the part beyond
the flank attacked, as A, for example, (*fig.* 32.)
and form on the line *e f*, before the assailing
army *c d* can overthrow, or entirely drive back
the flank B, on which its efforts are directed.
In this case, things are again brought upon
a par; for, the consequence would be a battle
front to front, the event of which is always
doubtful.

But if you face the line A B by corps detached for that purpose, as *g* and *h*, while with
the greater part of the troops you attack it
in flank, it becomes impossible for any part of
A B to throw itself into the line *e f*, till it
has overcome *g* and *h*, for which it is probable
there would not be sufficient time, if *c d* pushed on in a vigorous manner.   It follows that
the army A B, though the stronger, has now
no alternative but that of quitting the field of
battle, as it would otherwise be surrounded.
But, it should not have suffered things to go
so far; it should not have given the enemy
time to prepare: it should have acted offensively itself; for, in offensive operations, the
strongest has always great advantages.   In
its present situation, it should think only of

the means of making an eccentric retreat;
that is, to march its left wing to *i k*, and its
right to *o n*, if, *c d* does not stop its way
to the latter. The retreat of the right wing,
or of that part of the army nearest the flank
B to *o n*, should have in view to alarm the
enemy *c d* for his left flank *d*. It is by such
eccentric retreats, that a pursuit is prevented. The enemy would not dare to risk it, if
he wished not to be taken in flank himself, but
to avoid being exposed in turn to a concentric fire, and of course to terrible havoc.
Eccentric retreats in Tactics are as advantageous as in Strategics. Those of the latter
kind alarm the enemy for his lines of operation, and consequently prevent his advancing;
those of the former alarm him for his flanks
and rear, and prevent his pursuing.

From this we may conclude, that it is no
very great misfortune to an army, to be attacked in centre, and broken. It was one
among the ancients, but among the moderns
it is by no means attended with the like
dangers, contrary to Folard's assertion. When
that excellent writer published his system, the use of fire-arms in Tactics was
certainly not brought to its present perfection. Opinions were floating between the old

and the new system of war. In this unsettled state of things, the sagacity of Folard discovered defects, which gave him the notion of again assimilating our art of war to that of the ancients. He could not, at that time, know that this retrograde progression was impossible.

If the army A B (*fig.* 33.) is broken in the centre, it falls back eccentrically on *e* and *f*. By this movement, it puts a stop to all further progress on the part of the enemy *c d*, who broke the dotted line A B. For, if he advanced between *e* and *f*, he would be taken in flank on both sides, and be consequently obliged, to face both *e* and *f* at once. In this position, *e* and *f* might detach corps to the rear of *c d*, and act both against his supplies and his country. All that this would require, would be to send some of the troops from their flanks to the points A and B. They might even bear down themselves, by a flank march, on those points, provided they had, at *g* and *h*, magazines which would not be left exposed by that march, but would still remain protected from the enterprises of *c d*. There is yet a third manœuvre, which is, immediately to attack *c d*, who by his situation exposes both his flanks, as I have observed. In this ease, *c d* would have no alternative, but to act on the part of the flanks of *e* and of *f* opposite to the points A, B, in order to compel *e*, *f*, to fall back, and to form again in the direction of A B.

But in the present mode of waging war, nothing is more rare than this division of the two wings by an attack in the centre; because attacks of this kind are contrary to the spirit of the modern system.

A glance of the eye on figure 32 will be sufficient to convince us how preferable the attack in flank is. By this, the enemy is prevented from retreating so eccentrically as he may, when attacked in front. In fact, it is obvious that the army A B, being taken in flank by *c d* and compelled to retire, would find it next to an impossibility to fall back to *o n*, as *c d* would most probably stop its way thither. A retreat to *l m* would not be much more easy. A B, then, must retreat eccentrically towards A, or about *i k*, and there concentrate itself, in opposition to *c d*; but the two armies then finding themselves placed parallel to each other, and on the very spot where their flanks were before, the state of things would be equal on both sides: for, if A B, posted towards A, or *i k*, attempted, by a flank-march, to make a diversion in the country of *c d,* the latter would be able, by a similar march, to act against the country of **A B.** We have seen, on the contrary, that, after an attack on the centre, and an eccentric retreat which the army broken there would be obliged to make, the assailant, notwithstanding his success, is taken in flank on both sides, and is absolutely unable to attempt any enterprise against his adversary's country or supplies. Nay more, he is under the necessity in this case, as I have shown, to make a movement flankwise, either with his whole forces, or by detachments; and perhaps even to fall back, in order, to secure his flank, and to compel the enemy to resume a direction parallel to him. It is therefore evident, that an attack on the flanks of the enemy is more efficacious than one on his centre, and should consequently be preferred.

A defeat is of much less consequence in the modern system of war, than it was among the ancients. " The men that fly," says Folard, " are not dead." It may in reality be affirmed, that it is enough to think oneself not beaten, not to be so in fact. The number of the killed is never very great; that of the wounded is considerable, but they recover. It will be objected, that a great quantity of artillery is lost; but care is always taken, beforehand, to provide for that loss: and, indeed, the heavy pieces should always be sent into the rear, when the resolution to engage is taken, as they are of no use after the firing with small arms begins between the two armies. General Lloyd disputes the advantage supposed to be derived from the quantity of those pieces in a battle. He says, that six or eight pounders are as serviceable; and, at least, there is a possibility of removing them, and saving them easily.

Modern battles never weaken an army to such a degree, but that it may be ready for a fresh attack a few days after; and even that is not necessary, as every victory may be rendered fruitless by Strategic manœuvres on the flanks and rear of the enemy: but if the foiled army be the stronger, as I supposed, for example, A B (*fig.* 32.) attacked in flank and defeated by *c d*, then nothing is easier, immediately after a battle lost, than to manœuvre successfully an enemy weaker than yourself. Only forget that you have been beaten, put yourself again in a situation for acting offensively, and hem the enemy in on all sides ; this measure is infallible, the stronger having always an immense advantage in manœuvring. Although once defeated by the weaker, he will preserve his superiority ; but he must not be contented with defending himself; offensive war is his part.

It is not necessary to employ many troops to check the enemy in front, while his flanks are attacked. This may be performed in a more effectual manner, and with fewer men, by means of an open, rather than of a close corps. This mode of fighting has been allotted to the modern light infantry, to whom the French, in the war of the Revolution, gave the name of *Tirailleurs* *. The French owe their success in

---

* The author here lays the foundation of the system which he is about to establish relative to the *Tirailleurs*, and which I mean to controvert. No one is more convinced than I am of the great advantages that may be expected from them, provided they are employed according to the nature of the service to which they are adapted: but, I am far from thinking with the author, that it is proper, according to the modern system of war, to metamorphose the infantry of the line into light infantry, and to confine the instruction of the infantry, in Tactics, to the knowledge of deploying columns into the order of battle, as all other movements, according to him, may be executed as *Tirailleurs*. He seems, indeed, afterwards willing to modify his assertion, by saying that both kinds of infantry may be kept up in an army, exercising each of them in the peculiar service to which it is adapted. And here my opinion would accord perfectly with his, if he had not added, that the spirit of the new system of war required, that the number of light infantry should exceed that of the infantry of the line; a proposition which appears to me extremely questionable.

But, before I state the reasons on which I found an opinion different from the author's, I think it necessary to ob-

that war to their adoption of this method. But, in my opinion, the manner in which the light

serve, that the notion he has formed of *Tirailleurs,* if we may judge from the system he attempts to establish, is far from being conformable to the manner in which they have been employed by the French, with whom he has probably never waged war; and that, as he asserts that the latter owe their success to them, it is natural to imagine that he would not have deviated from their method so considerably as he has done, allowing that he did not choose to conform to them strictly.   I shall briefly state the origin of the system of *Tirailleurs* in the French armies, and the mode of their employment in them.

At the commencement of the war brought on by the Revolution, the armies of France were composed of troops of the line, instructed, it is true, but entirely disorganized by the Revolution, and of battalions of volunteers, which, levied in haste, and commanded by officers almost all strangers to manœuvring, could have but a very circumscribed degree of instruction and organization.   The French were sensible how disadvantageous it would be to engage, with such troops, in regular actions, and conformably to the Tactics then in use, against armies perfectly disciplined and skilled in manœuvring; they, therefore, sought for a new system, which, while it enabled them to take advantage of the character of the nation that was peculiarly adapted to the shock, of the personal bravery of the soldiers, their enthusiasm, and their superiority in number, removed, at the same time, the dangers to which regular manœuvres, badly executed, would have exposed them; in short, they adopted the system of *Tirailleurs,* and the *attack in columns;* and firing, became with them merely a preparatory movement.

infantry are instructed is not a good one ; it is still too stiff, and does not appear to me to an-

By means of an artillery, as numerous as excellent, they soon awed the enemy's batteries, and a cloud of *Tirailleurs,* occupying the whole front of his line and keeping up a brisk fire, prevented him from seeing on what point the real attack would be made. If these *Tirailleurs* observed any irresolution in the part of the enemy's line opposite to them, they rapidly formed into columns, and advanced briskly to charge that part, which they often drove in. At the same moment other columns, entirely formed of infantry of the line, hastened to the points broken by the *Tirailleurs,* and completed the affair.

If they had to attack the enemy in a position strong by nature, that is to say, by the difficulties which the ground presented to a regular attack, the *Tirailleurs* took advantage of spots through which they might creep, to approach the position as near as possible, in order to reconnoitre it ; and as soon as this preparatory measure was concluded, the French, considering the position as the front of a fortification, took the prolongations most advantageous to them, and advanced in the direction of the capitals their columns of attack, whose passes and march they took care to secure by a number of fixed batteries. At the same time, they erected strong ones opposite all the points whence the Austrians could come out on the flanks of the columns, and carrying a great force on those which they intended to attack, they contented themselves with alarming the rest of the line. Their artillery thus disposed, served, in the first instance, as a powerful support to the attack, and if this succeeded, once that a column had cut through and taken in rear a portion of the line of the Austrians, the latter had only to

swer the end so well as another method, of which
I shall presently speak. That which is in use

think of their retreat, for there was no longer time for
manœuvring. If, on the contrary, the French columns
were repulsed, then their artillery covered their retreat, and
they returned quietly to their former position, only with
the loss of some men, whom it was extremely easy for them
to replace; while the Austrians, necessarily frugal of theirs,
from not having the same means to recruit, and too much
constrained to preserve their position, did not dare to come
out of it to pursue the French; and turn their advantage to
account. The French, therefore, were very rarely entirely
worsted in an engagement; for, besides that they never
really brought more than a part of their line into action,
they had it likewise in their power, as I have said before,
to repair their losses immediately, and in a short time to re-
new their attacks, by means of which they at length succeeded in breaking the Austrians, and forcing them to retire.

I will not here inquire what was done at that time, nor
what might have been done, to oppose with effect the French
mode of fighting. All digression on this head seems to me
the more useless, as, besides that the events are past, the
French, in my opinion, designed, in the adoption of that
mode, rather to create a system for circumstances, that
is to say, such as their situation at that time required, than
to effect a revolution in Tactics, and establish a new permanent system. Is not the present organization of their infantry a convincing proof of what I advance? The light
infantry forms scarcely a fourth part of it: and why should
the French, who, according to the author himself, owed
their success to the *Tirailleurs,* have adopted their present

is as follows : The men, drawn up in two ranks,
are divided in such a manner as to leave a space

organization? Why should they have resumed a regular
system of Tactics if they had not themselves been sensible of
the advantages of it, and felt the inadequacy of a system,
which circumstances indeed had rendered necessary for them
for a certain time, but which they dropped as soon as they
could? Besides, were not the French *Tirailleurs* exercised,
not only in forming columns, as the author requires, but in
charging in the same order? Were not their movements
supported by their numerous infantry of the line, whose excellent instruction, at first paralyzed in consequence of the
temporary disorganization produced in it by the Revolution,
had resumed all its effect after the re-establishment of its
disicipline? Was not their cavalry likewise restored to their
former state? In fine, had they not a numerous and excellent artillery, a great superiority of numbers, or at least
the power of repairing their losses with ease and expedition, which the Austrians could not do? Such are the principal causes of the successes of the French, and not particularly their *Tirailleurs*, who would never have procured
them the advantages they obtained, without the concurrence of their infantry of the line, their cavalry, artillery,
and, generally, all the means of which I have been speaking.

The French will continue, no doubt, to employ *Tirailleurs*
and columns of attack : this kind of fighting agrees and will
agree with the French impetuosity, as long as the national
character continues what it is. But, besides that the system, taken altogether, is very different from that which the
author wishes to establish, I take upon me to predict that
the French will not confine themselves to that system, and

I

from man to man, and from rank to rank.
(*fig.* 34.) The soldiers in the second rank,

even, when they do use it, they will modify it as much as
their present situation and circumstances may require: their
light infantry, though numerous, will probably be always,
as at present, much less considerable in number than that of
the line, which, contrary to the author's system, they will
take care not to give up, or even decrease. In short, they
will make use of their *Tirailleurs* in the service to which
alone they are adapted ; that is to say, they will, to make use
of a general expression, employ them in preparing operations
to be carried into execution by the regular corps.

The author admits too, that his system requires a numerous cavalry; and, truly, it must be very numerous to protect
an immense quantity of scattered soldiers, who, especially
in a level country, could not support themselves without its
assistance. But, has he well considered the disadvantages
that would attend a cavalry, the numerical force of which
should exceed the bounds of a certain reasonable proportion,
which, according to the opinion of the most experienced military men, it ought to maintain with the rest of the army?
Would not such a cavalry, instead of easing and facilitating
the movements of armies, be extremely puzzling in most
countries to move? What difficulty, besides, would there
be to subsist it? What enormous magazines, what a chain
of convoys, what immense communications, would it not require to be kept up, if the army march to a certain distance? Would-it not frequently happen that the most important views of the war would be relinquished from the
dread of being at a loss for subsistence? It will be no answer to say, that this cavalry may forage : resources of that
kind are too precarious to be prudently trusted to. I will

placed facing the intervals left between those in
the first, protect their flanks.   When advancing,
the second rank C D, passing through the intervals

not add to what I have said, observations on the immense
expence that would be necessary for the purchase and re-
placing of the horses; on the difficulty of procuring them,
&c.

It appears to me, then, that the author's system is inad-
missible in practice in every respect, which I could have demonstrated still more circumstantially, had I thought it necessary to discuss it more at large, and detail all the objections
which I could collect against it.   I will only observe further, that we shall find him saying himself, in the 1st chapter of the 3d part of his work, " that it is impossible, let
what will happen, ever to do entirely without infantry of
the line," which proves that he feels the necessity of establishing, at least, the modification he proposes, that is, of
having at the same time both infantry of the line and light
infantry in an army, provided the latter be more numerous
than the former.   Of this modification too I have already
stated my opinion, and shall conclude this note with saying,
that convinced as I am of the utility of *Tirailleurs*, when
they are confined to the kind of service to which they are
adapted, I do not dispute the necessity of a numerous body
of light infantry in an army; but only affirm that their
number should be inferior to that of the line, which ought
always to be the basis; and that the army whose manœuvres and movements are made with the greatest unity and
ensemble, will ever have an indubitable advantage, allowing
all other things to be equal, over that which shall possess
those qualities in a less eminent degree.

COMMENTATOR.

of the first A B, move up to the line E F, and fire. To retire, they pass again through the same intervals and return to the line C D. The great advantage resulting from this is, in the first place, that of forming a front of greater extent than when the men are close; in the second place, a more destructive fire is kept up, for it never ceases, and each soldier takes a better aim from being more at ease; and, in the third place, a less number of the troops are killed, as many of the enemy's balls are lost in the intervals. But in the method, I propose, all these advantages are more eminently united. The soldiers, after they are scattered, are not to move in a right line, but circularly, in the following manner: As each of the *Tirailleurs* composing the first rank fires, he faces left about (*fig.* 35.) and runs circularly to the place in which the man of the second rank behind him stood. The latter advances rapidly, also circularly, exactly to the same point where the former was, and fires while the man now behind him reloads to return and fire again. The cavalry, in this method, are posted on the flanks. All is incessantly moving, the enemy is perplexed by it, and the soldiers, never stopping or moving in a right line, are much less exposed to the balls of the enemy than in the other method. This, however, is only good while the enemy keeps his ground. If he flies, the method of advancing directly and of keeping up a continual fire, as shown by figure 34, is certainly the most effectual. If you retreat yourself, then you may manœuvre as is shown by figure 36; that is to say, that each *Tirailleur*, after firing, instead of returning to the spot where the man placed behind him stood, shall move lower, as is seen in No. 3, of this figure. In this manner, by insensibly curtailing the arch, you may dispute the ground foot to foot, and only yield it in the midst of a fire constantly kept up, which will check the enemy, and ward the shock of his impetuosity.

Still, this mode of retreat possesses no other advantage over that in a right line, than the continual motion of the troops, by which they are less exposed to the enemy's fire. It requires, however, fewer men than the latter manœuvre, as the soldiers may be more scattered.

If I am told that the fire of a close line is superior to that of a body thus scattered, I answer, that I here speak only of checking the front of the enemy, which is but accessary to the real attack directed against his flanks: for, I maintain, that it is only the flanks that ought to be vigorously attacked, and that a

I 3

general should content himself with amusing
the front. Now, irregular firing is excellent for
this purpose. If it kills few, it wounds many :
only it is indispensable that the *Tirailleurs* be
not at too great a distance from the enemy;
the troops must be bold, and must not begin
firing, when he is scarcely to be perceived on
the horizon. If it be objected, that it will be
easy for cavalry to cut to pieces a body of scattered *Tirailleurs*; I reply, that these, on their
side, must be supported by cavalry, placed in
a second line behind them, and that, in case of
a retreat, they should take refuge very quickly
among their cavalry, where they will be safe.
Every thing considered, I should wish the
retreats of the *Tirailleurs* to be made *à la
débandade*, that is, in a scattered and irregular
manner; not slowly, and firing as in the methods
above described. It is enough that the cavalry is at hand to cover them; but they should
return to action with the same promptitude. All
these movements are entirely conformable to
the spirit of the modern art of war. If the troops
are pursued by the enemy's cavalry in an open
level country, an action between the mounted
troops must take place, in which it is the duty
of the *Tirailleurs* to support their own. If there
is a wood near, if the country is intersected by

ditches and hedges, the cavalry becomes altogether, of no use in protecting the light infantry; but then the latter, in such a country, is sufficiently secured from the attacks of the enemy's cavalry: they begin firing again from behind those natural entrenchments, and are in safety while they destroy. The front of a hostile army is much better checked in this manner than by close corps and cannonades.

It is equally necessary to draw the enemy's attention, when, not meaning to attack a flank, it is merely intended to manœuvre with a view of covering, or gaining a march, and of acting against the supplies of the enemy.

It will still perhaps be said, that it is better, in an attack on the flanks, to advance in close ranks, because the lines of fire, being more compact, yield a better sustained fire: but I reply, that once on the flanks of the enemy, and near enough to use the firelock, it is a matter of indifference whether the attack be made in close ranks, or as *Tirailleurs;* for, be it how it will, the enemy, if vigorously attacked, will be beaten. Nay, the assailants, when formed as *Tirailleurs,* have more lines of fire than the enemy can oppose to them, as they envelop him. In such a position it will be difficult for him to form on the line *e f* (*fig. 37.*) espe-

I 4

cially if he is checked in front, which should be the case; and, if as I have said, there is cavalry at hand to support the attack.

Besides, if it be imagined that the fire of a platoon, or of a battalion in close ranks, will produce more effect than the irregular fire of a scattered corps, because, from the whole firing together, the enemy receives at once a greater number of balls; it must be from want of reflecting on the fact testified by every soldier who has seen actual service, namely, that when infantry are in action, the firing by files of itself takes place*. But then this kind of fire

---

* The author is certainly in the right, and the only regular fire I think generally possible in war, is that by battalion or demi-battalion. As for that by platoons, which, on the authority of the king of Prussia himself would be unquestionably the best, *if it were practicable,* it appears to me, as well as that by divisions, to be suitable only to infantry posted, who want to keep off or check attacks, and particularly when covered by intrenchments: for, independently of the noise, which may occasion many mistakes, it is, I may say, impossible to give soldiers, in presence of the enemy, a sufficient degree of coolness to attend to the word of command, and to observe that caution which regular firing by platoons and divisions requires. Nor am I very certain that it is any great misfortune to an army, that, after a few discharges, the fire by files takes place of itself. This fire being at will, a soldier, notwithstanding the smoke, can, in my opinion, aim much better than he is able to do when

in close ranks is certainly not equal to that of
soldiers standing wide, who, not pressing upon
one another, have unquestionably a freer use of
their bodily powers, and may aim their shot
with much greater exactness. Besides, infantry
have a mechanical manner of handling the firelock, in consequence of which they can hit only
at a certain distance; but a soldier not in the
ranks is at ease. In short, it is well known that
the greater part of the balls from a platoon, or
battalion fire, pass over the heads of the enemy,
and kill very few men*.

I shall perhaps be reminded here, that I have

firing at the word of command given by the officer, who
frequently, through excessive eagerness, hurries him, and
obliges him to fire before he has had time to aim properly.
Other military men no doubt, as well as myself, must have
frequently remarked this inconvenience: hence then it follows that the fire by files, which besides is very brisk, must
do more real execution than a regular fire; that is, however, if each soldier does his duty, and has been taught to
aim by a less imperfect method than the soldiers generally
are. This fire has the further advantage of animating the
men, keeping them constantly in action, and rendering them
insensible of danger: but they should be accustomed to
cease firing the instant the signal is given. I shall resume
the subject of firing in the next note.—COMMENTATOR.

* Sometimes too the soldier fires too low, and the reason
of this inaccuracy, in either case, is that the soldier, accustomed, as the author says, to a mechanical manner of handling
his firelock, most frequently levels it at the same height,

defined the movements of Tactics, to be those made by an army while actually manœuvring, and ready to engage the enemy, or to repulse his attacks; whence it must follow, that as the *Tirailleurs* are not restrained to regular movements, their mode of fighting ought not, according

withojn attending sufficiently to the difference of the distances at which he stands from the enemy, or to that of the positions he occupies with regard to him. The best way of remedying this fault would be, no doubt, not only to exercise the troops frequently in firing at a target, but to vary the distances, and the nature of the ground, placing the target sometimes on a sloping ground, sometimes on an eminence, varying also the slope and the height. By these means, the soldier would learn how much, according to circumstances, he should raise or lower his aim. He should, likewise, accustom himself to judge of distances by his eye, as on them must depend, in a great measure, his manner of aiming; and, in my opinion, he will not find it very difficult; for, besides that those which it would be necessary for him to judge of, would not differ much from one another, and that being exercised as I have mentioned, he would, by daily observations, acquire the advantage of practice, which is the best of all masters, those distances too, would never be very long. In fact, though the distance to which the musket carries, that, I mean, at which an effectual aim may be taken, is in general supposed to be from two hundred and forty to two hundred and eighty yards, I am, notwithstanding, convinced that, to do real execution, infantry ought not to fire further from the object, than from one hundred and sixty to one hundred and eighty yards at most.

COMMENTATOR.

to my definition, to be classed under Tactics.
I have several answers to make to this observation. In the first place, the troops always put
themselves in motion tactically, that is to say,
in close ranks; from this line of battle, which
presents a straight and close front, they scatter
and attack as *Tirailleurs;* secondly, they ought
to be exercised to reunite promptly, and resume
the form of a regular corps when the case requires it, as, for instance, when the cavalry attacks them; thirdly, attacks *à la débandade*
must be such only to the eye, but, in fact,
union and order must prevail in them : thus,
we discover, in these sorts of attack, marks that
denote Tactics. I own, however, that I am not
satisfied myself with these answers to the objection. In fact, it is possible, no doubt, to advance
upon the enemy in an irregular and scattered
manner, so as not only to come within musket-
shot, but even to intermix with his troops, as
the Turks do, and certainly this mode of
slaughtering must not be styled an operation of
Tactics; but it is equally out of the sphere of
Strategics. In this case, I must take shelter under
the definition formerly laid down as the most general ; that all operations made within sight of
the enemy's army, are operations of Tactics.

This scattered mode of fighting is particularly
suited to new-raised and undisciplined infantry.

It is a rule to avoid engaging, and particularly in general battles, when a commander has only new soldiers to oppose to veteran troops, and to be contented with manœuvring; as, for example, with attempting to cut off supplies, to attack magazines, and to make diversions in the enemy's country on his flanks and rear: but during these operations the commander must check his adversary's front, which cannot be more effectually done than in the manner I am going to describe, and which does not require so much order as the engagements of light infantry, of which I spoke above. It is sufficient to teach the troops to take advantage of the nature of the country. The soldier must be taught to creep along the ground up to the enemy, to fire and load lying down. He must learn to shelter himself by trees, and to throw himself into ditches or behind hedges, so as to fire without being seen, and to prevent the possibility of being hit by the enemy's balls. These are exercises which young people will easily learn. Still, when time permits it, it is of importance to train them to manœuvre as a regular corps: and let an army be ever so newly raised, it must at least know enough to deploy from column into line. It is also necessary that it be divided into companies, battalions, and so on, and have a sufficient number of officers, for it would be impossible without all this elementary organization to lead it. With respect to cavalry, it is indispensable that it should be still more trained, that it should be able to make a regular attack in close ranks in the most rapid movement; for its power consists chiefly in its being compact, and in its velocity, by which it must easily overcome an infantry without depth. Lastly, the cavalry ought to be numerous; particularly if the seat of war be in an open and level country.

The retreats of the infantry intended to check the front A B (*fig.* 38.) ought not to be eccentric, nor aside, because the main object in view is to divert the attention of the army A B from his flanks which are meant to be attacked; but these retrograde movements are to be made directly towards *l f*. If it be a serious retreat, if it be intended really to leave the front A B, and prevent a pursuit by alarming it for its flanks, then the retreat should be made eccentrically towards *g h*.

I have already shown that an army should never attack the flanks of the enemy, without at the same time checking his front: otherwise, he would probably not fail to throw himself into the parallel line which I have marked *e f*: (*fig.* 31. 37.) Of course, all the advantage of the attack in flank would be lost, and things would

again become equal on both sides.   I have also
said, that if the front of the enemy be checked
by a close corps, it would not be so efficacious
as making use of light infantry for the purpose,
supported, according to the nature of the ground,
by cavalry.   It would however be infinitely
better than making an attack on the flanks of
an army, in an oblique position like that shown
in figure 39.

For, in taking this kind of position, which was
invented by Epaminondas, investigated in mo-
dern˙ times by the Chevalier Folard, and made
use of by Frederic II, an army exposes its own
flank in case of a retreat, as is shown by the
dotted line D e.   It would likewise be very easy
for A B, to come upon the flank D, in stretching
into a right line by a side movement towards f.
The army A B might also bear upon the line
A g, by which it would not only avert the attack in flank which C D has undertaken against
it, but might, likewise, by stretching, out flank
the wings of its adversary.   Indeed, A B may
not always be able to bear upon the line A g,
particularly if the enemy be very near, and if
he has much cavalry.   For, the cavalry might
overtake the infantry of A B, and compel it to
stop before it has executed its movement ; or if
it is opposed by a corps of cavalry, it would

probably join and route the latter which retreats, before it could form in the line A g. But in all cases A B, by its right wing A, has it in its power to assume the offensive against the left wing of C, and to reach its flank by an expeditious march, as the column h i shows. What would be the consequence? If C D take A B in its flank B, A B will in its turn take C D in his flank C. The two parts attacked will probably be beat by the parties attacking; and after the engagement, the two armies will remain opposite to each other, and be placed in a direction rather oblique to their former front. Such must be the result of all these movements, if A B acts as it ought to do. Add to this, that the left flank B of A B would not even be beaten, if it fell back quickly to the line A g.

It is impossible, then, to attempt an attack with a close oblique line as C D, if the army be not previously completely upon the flank of the enemy; and if, for example, the flank D is not at least at the point f on the prolongation of the front A B. I do not know if even that would be enough. It would be necessary to be sufficiently advanced on the rear of the enemy to overturn him, and without that, success is not to be expected, unless the enemy be supposed to be very unskilful. It is indispensable, likewise, that the front of the adversary be checked, and particularly the wing opposite to that attacked, according to the example set by Epaminondas at Mantinea I have already explained the manner of performing this manœuvre to obtain the best success.

It is not necessary to follow the latest mode in Tactics, for throwing back a wing into the line A g, so as to prevent its being taken in flank. This mode may in general be given up in all changings of front, as they may be executed more briefly. It suffices to make the platoon, which is the pivot, turn their shoulders towards the new direction to be taken, and to direct insensibly the others backward to the same line, without breaking the front. If a certain extent of ground is to be run over, it will, no doubt, be necessary to face right about; but the end will be sooner obtained by this manœuvre, than by the method of wheeling. I have seen raw troops changing front in the twinkling of an eye, in consequence of the mode I recommend. The platoons not having occasion to remain close to each other, may run. It would even be possible by this method to take a position entirely back, and so as to let the two wings project from their angles. All that would be requisite for this, would be to give the new point of direction, and

then to let the manœuvre be performed almost
à la débandade, and without observing any other
order than that each man should take care not
to separate from his neighbour the line would
be more speedily formed again than by the
method usually practised.   Infantry exercised
in breaking their ranks, and then speedily forming in line again, would, if led by commanders
who knew how to take advantage of the ground,
envelop the flanks of an enemy advancing upon
them in close ranks, before he could make
proper dispositions to defend himself.   A close
line can advance but slowly: the Prussians have
settled its march at seventy-six steps in a minute
It cannot keep the line, or even avoid opening,
if it moves faster.   Such a line, then, can proceed but heavily, while on the contrary the light
infantry may put their legs to the utmost speed,
provided the soldiers have been trained to find
their places again without difficulty.   They need
only stand two men deep; for, a line of fire should
not proceed from three ranks: these are too few
when the object is to break the enemy by shock,
so as to produce a decisive effect, and too many
when it is merely to fire, every thing then depending on extension.

I assert that it is not even necessary, that the
infantry, after scattering, should always form

K

again in ranks. Let us suppose, for example, that
A B. (*fig.* 40.) is attacked by the line **C D**; the
left wing might scatter and run towards *e f*, and
there facing right about, at a certain signal, return with a little loss of breath, and, without
forming into ranks, attack the left flank **D**, which
it would salute with a rolling fire on three sides,
before **D**, to defend itself, could have had time to
take the form of a hatchet **D** *g*. But, that such an
attack may be successful, the enemy's cavalry
must not be near, or must be checked by yours:
if you had any thing to apprehend from them,
the precaution to be taken would be to form
into columns, of which I shall by and by speak
more at length.

If attacks and retreats, then, be made according to the manner here proposed, and if, above
all, care be taken to have them supported and
covered by a numerous cavalry, the greater part
of the evolutions of the infantry would be at an
end. It would, however, be always indispensable, and I have already said so, that the troops
should know how to deploy into line: in the
changings of front, this manœuvre might be
made by letting the corps file off running, by
platoons, as I said before: there would even
be no inconvenience in making use of this
method for all the side movements to be made

while formed in line; but it would not suit route marches. In fact, it is as important not to range the columns too deep, as to extend the front when troops are deployed. Of what a length would the columns be, if they marched only two or three men abreast? What a time would it take to deploy? For the deeper the columns, the longer is the time which their deployment requires. It is to be observed, then, that as large a front as possible is to be given to the columns, as five or six men for example, if the ground will permit*. Josephus tells us, that the Romans made their sections six files : the Prussians make them five.

In a flank, or parallel march, the troops form into line by means of a single quarter wheel. This movement may be made the more rapidly, that the divisions have but to march few paces to uncover one another; therefore, this manner of marching should not be a matter of compulsion, but chosen as often as possible.

---

* It is proper, if the ground permits, that the fronts of the ROUTE columns should be even larger than the author advises. A front, from twelve to sixteen men, requiring the road in which the column marches to be only from *twenty to five and twenty paces* broad, is by no means too large, provided, I say, that the nature of the ground is suitable.—COMMENTATOR.

On the contrary, in a march perpendicular to the enemy's front, it is necessary, in order to form into line, to deploy the columns, which requires infinitely more time; for, the divisions, to reach their place in the line of battle, have to move along either the two smaller sides, or the hypothenuse of a right-angled triangle. If the deployment is regularly made, according to the Prussian method (*fig*.41.) the divisions, in their movement, march along the two smaller sides of that triangle *.

In fact, let us suppose that the column A B, with the right in front, should deploy into line in the alignement A C on the rear division B; this division would indeed have only the distance A B to pass, in order, when it is uncovered by those before it, and which are regulated by it, to come up to the point of formation A, that is, to the very ground which the front division has quitted; but the other divisions, except that of No. 1, which being already on the alignement A C, has but to march strictly in that alignement to come up to the point of

* The mode of deploying, of which the author here speaks, is that which the French call *deployement à tiroir*, and is adopted in almost all armies. See *Rules and Regulations for the Formation, Field Exercise, and Movements of his Majesty's Forces*, page 182.—COMMENTATOR.

formation C which its flank is to join, should each move, first in a parallel direction to A C, in order to uncover the division behind it, and afterwards perpendicularly to it, to reach the place which it is to occupy on the new line.

If a column (*fig.* 43.) with the left in front, should deploy into line in the alignement A B, on the division A, and if, for greater dispatch in the deployment, the other divisions, after facing to the right, marched diagonally to the line, they would each wheel twice; first, by files, to bring up their right flank to the line, when at a distance equal to their front, from it; the second time, to form in line, when all the divisions having their right flank upon A B, the command is given for a general quarter wheel.

Captain Rösh, a Prussian officer, proposed a mode of deploying, which appears to be the most easy, and of course the best: the platoons, in marching towards the alignement A B, (*fig.* 42.) keep between each other a distance equal to their front. As soon as the platoon 1 comes into the alignement, the command is given RIGHT or LEFT, FACE, according to the side on which the column is to march; the next platoon marches without altering its step, up to the identical point where the preceding one has faced to the RIGHT or LEFT, and does the like; the third, fourth, and all the others follow this example. Each platoon having thus gone over its distance, comes upon the alignement, that which marches immediately before having already made room for it; by this means, one of the wheelings occasioned by the preceding method (*fig.* 43.) is avoided; for, the line is then formed by a simple HALT, FRONT; and wheeling is a movement that always requires a certain length of time, as it is executed on the arch of a circle *.

---

* I own I cannot see any advantage in the method of deploying (*fig.* 42.) proposed by Captain Rösh. In the first place, as to the time: for, if I compare this method with that *à tiroir* (*fig.* 41.) I find that in both, the divisions have generally equal spaces to go over, and the same movements to make, to reach the place which they are to occupy on the new line; excepting the case, however, where, in the deployment *à tiroir*, some divisions should meet in their march obstacles which would oblige them to manœuvre, in order to avoid them. If I afterwards compare this method with that represented by figure 43, I find, indeed, as the author says, that the latter requires one wheeling more than the former: but I also think that the time required for that wheeling is more than compensated by that gained by the divisions which are detached from the column, in marching diagonally to the new line.

As to the greater facility which the author ascribes to Captain Rösh's method, compared with the two others,

A method of deployment, where the divisions move only along the hypothenuse of a right-angled triangle, has been introduced among the Prussians: they call it *the adjutant's march.* These adjutants, who, from habit, know the length of the front of their battalions, measure, by galloping their horses on the alignement, the space necessary for their forming on it into line. (*fig.* 44.) Each battalion quits the column, and marches by the straightest road towards its adjutants, who have placed themselves at the points of APPUI, and intermediate points, marked by the numbers 1, 2, 3, 4. As soon as the head of the battalion comes up close to its adjutant, it deploys according to the method already described. If the officers commissioned to trace out the front are not

(*fig.* 41 *and* 43.) I am as much at a loss to see on what he founds his preference; for, this method does not require less attention than those in preserving the distances, which is the chief thing; and it does not seem to me to furnish much easier means of observing them. Indeed the uniform and continued movement of the divisions, marching with an equal step, and in the same directions, may give some facility: but, even allowing this advantage to Captain Rösh's method, it would not be the less subject, in other respects, to inconveniences, which, as every officer will doubtless perceive them, I think it needless to explain; for, it would lead me unnecessarily into a very long digression.—COMMENTATOR.

K 4

very erroneous, there is no doubt that the battalions will more speedily come up to the alignement by this, than by the preceding method,
the hypothenuse being shorter than the two
other sides*.

And here is the place for saying something of
points of MARCH, and points of APPUI, towards which the troops march when they advance. None of these points were marked out
during the seven years war, in the numerous battles which signalized that period. Nothing, in fact,
is easier, when you see the enemy, than to march
to him, so as not to present him a flank, and
to keep yourself always nearly parallel with
him. Indeed, the difference which there may
be between the directions of two hostile armies, scarcely deserves to be calculated at so
great a distance, as the troops always begin to
march in order of battle at several thousand
paces off, and out of cannon-reach. Besides,
it would be necessary, if such great importance
were attached to the points mentioned before,
to mark out others at all the movements of the
enemy, which are frequently very quick. I

---

* The author here alludes to the deployment in mass of
battalions. See page 301 of *Rules and Regulations, &c.*
which I have already cited.—COMMENTATOR,

am persuaded, that the flank may be secured
from the attack of the enemy, without so many
precautions. It is sufficient to see him, to
be able to take, very speedily, a position
nearly parallel to his; I say nearly, for, in this
case, it is as impossible to attain mathematical precision, as it would be ridiculous to require it.

It is very practicable, then, to do without
those points *; and, when an army marches in
line, the chief thing (after the enemy has been
sufficiently examined) is, to be guided by the nature of the ground, and of the advantages it
may give; from which, and from the position

* I agree with the author, that, at the distance at which
an army generally begins to march in order of battle, a
little more or a little less regularity in the dressing of the
line, is not of very great importance: as this dress, however, will probably have a great effect on that to be observed in marching, and as any considerable irregularity,
when nearer the enemy, might be attended with serious
consequences, I cannot but think that the points of
MARCH, and the points of APPUI, are useful; especially
when the line is of great extent, and the ground it has to
go over not very level. Indeed, I am far from imagining,
to use the author's expression, that mathematical precision is necessary, nor even such as might be looked for [in
the evolutions of field-days in time of peace; yet it seems
to me that, by totally renouncing those points, the line would
be deprived of a great means of directing itself with faci-

of the enemy it is, that the points proper to be occupied, and those that ought to be neglected, are afterwards to be determined. Nor, indeed, is it by any means necessary; as I have proved, to form a line, the points of which should be all close to one another*.

lity, and that to restore indispensable order in the march, it would be necessary to execute partial manœuvres, which, besides that it would fatigue the troops, must slacken the movement. In support of my opinion, I shall adduce the practice of the Prussian army, which, beyond contradiction, marches in a superior style, and which, at the time I am writing, not only makes use of the points mentioned, but takes greater precaution, perhaps, than any other army, to have them accurately attended to.—COMMENTATOR.

* We may conclude, from what the author says, without his acknowledging it formally, that he is no advocate for those orders of battle which are fully determined beforehand, and sometimes the preceding day, on reconnoitring the position of the enemy and his apparent arrangements. He is very right; for how many armies have been defeated from this method being followed! Is it not in the enemy's power to alter his arrangements before the time of action, and thus frustrate the plan previously formed, however well combined it might have been, in respect to his situation at the time. It is, therefore, only in sight of his adversary that an able general finally settles his arrangements. He keeps his army in columns, by which he has it, as it were, in hand, and retains the power not only of guiding it rapidly, but also of making interior movements which escape the enemy's notice, or mislead him. If the general determines

At present, the deployments of columns are executed at a distance from the enemy, or, at least, out of reach of his artillery, and are

to be the assailant, he reconnoitres from the front of his vanguard the situation and dispositions of the enemy : if neither one nor the other offers a weak part, against which he may advance rapidly and advantageously, he begins to manœuvre in presence of his adversary, and endeavours, by misleading him, to induce him to make some wrong movements which commit him, and of which he instantly avails himself. If, on the other hand, not being able to mislead the enemy, or make him commit a fault, the general judges that he cannot attack him but with disadvantage, he falls back, unless reasons of the greatest weight determines him otherwise, without having come to any engagement, and takes a position, where he waits a more favourable opportunity.

Supposing the general now to be, on the contrary, assailed by the enemy ; he does not open his defensive dispositions till he is clearly acquainted with the points which the attacking army proposes to attempt, and keeping till then his army in columns on the field of battle which it is to occupy, he determines the arrangement of the troops according to the movements of the enemy.   The latter perceives, at the principal points of the line he means to attack, only heads of columns of which he cannot calculate the depth, or divine the object, and finds himself at a great loss. If he manœuvres, the army manœuvres also ; if he endeavours to mislead, so does the general, either by presenting him a point weak in appearance, in order to attract him towards that point, where, according to measures skilfully concerted, he will rapidly collect a force to defend it, or, by inducing him to make a wrong movement, which affords an opportunity

covered by a strong van-guard, drawn up
in line. In plains, it is the part of the cavalry to cover this movement, and the march
in line which follows it: with this precaution,
the operation is safely effected, and does not
require such great tactical precision as is usually imagined. But, I repeat, that this is the only
military operation in which it is indispensable
that infantry should act conformably to the regular Tactics in close rank: all others may be performed by *Tirailleurs*. But, as these deployments
of columns should never be attempted except
when secured against the attack of the enemy,
they do not, of course, require so much exactness,
as movements which take place during the time
of actual fighting; it therefore follows, that
troops confined to this manœuvre alone, have

to attack him to advantage, and to make immediately an
offensive counter movement.

It is evident that what I have said is applicable only to
positions that are open to attack on several points: for, if a
defensive position, for example, is so advantageous that the
ground necessarily reduces the attack to one point, there is
then certainly no inconvenience in determining beforehand
your order of battle; because there can be no uncertainty
respecting the point where the greatest force must act, and
because the knowledge which the enemy has of the position
he occupies, will, undoubtedly, prevent his being led into
an error.—COMMENTATOR.

no occasion to be so well exercised, as those intended to fight as regular and compact infantry.   This mode of manœuvring is, of all, the
most difficult to learn, and requires the longest
time for training the soldiers.   It is also indispensable to teach *Tirailleurs* to load and fire in
a manner different from that in which it is commonly done.   In the usual mode, the soldier
does not fire as an individual, but as part of a
whole; as a member of a collective body.   As
a *Tirailleur*, he must know how to make use of
his gun as if he were by himself; and not
being cramped by his neighbour, but able to
act freely, he aims, and his shots tell.   I have
already said that he ought to be able to fire and
load lying down, to take advantage of every
state of the ground, and to conduct himself
like a man of sense, who expects nothing of
others, and depends upon himself alone.   I do
not deny the possibility of having an infantry
qualified for both the kinds of fighting of which
we have been speaking: the only question is,
whether it be possible to attain perfection in
both at the same time? and whether, if one
must be chosen, the *Tirailleur's art* is not to be
preferred?   It is much easier to be taught to a
multitude of new men, than the Tactics of the
line, and of regular corps: and, as it is prac-

ticable to do with that alone, provided the *Tirailleurs* be supported in open places by a numerous cavalry, would it not be right to give
the troops only that kind of instruction, and to
neglect the other, except, let me repeat, what
respects the deployment of the columns? Besides, it does not require so much time as is
commonly thought, to form a troop of infantry,
even to the Tactics called *heavy*; for we have
printed Instructions, by which a man may be
made a soldier in forty days; but if the thought
of making the same men fully perfect in both
kinds must be given up, it would be well to
divide armies into *heavy* infantry and light infantry; to exercise them only in what is respectively proper for them, and to make them
at least equally numerous: indeed, the new
system of war even requires that the prepon-
derance of numbers should be in favour of the
light infantry. As to the cavalry, I have already said that, to be of any use, it is absolutely necessary for it to keep in a body before
the enemy.

It is still a question, whether infantry, armed
as they are at present, are able to resist cavalry. It appears to be demonstrated, that
the musket and bayonet do not secure a column of infantry, even formed upon Folard's

system, and which defends itself with the greatest coolness, from the danger of being broken and cut to pieces by a body of horse, equally brave, let its position be ever so advantageous. The History of the War of the French Revolution, furnishes an example of it: when Prince Hohenloe, after joining the Prussians, beat the French near Kayserlautern, three battalions of the latter formed into a column so close at the moment when they found themselves threatened by the Prussian cavalry, that there remained no space between the divisions. Katt's regiment of Prussian dragoons charged this column: it defended itself so bravely, that the dragoons were obliged to cut their way with their swords; yet it was soon annihilated.

This instance proves the Chevalier Folard to be right in giving his column of infantry pikes, mixed with the guns and bayonets, so that the long weapons might be supported by the short ones.

If a column is exposed to such a fate, how would it be with an infantry drawn up two or three deep? The advocates for fire will say, that a battalion deployed, sends more balls against a cavalry which charges it, than a column can. But this is only true when a small front of cavalry charges a large front of infantry, which keeps up against the troopers oblique and cross fires. If the front of the cavalry is equal to that of the infantry, the oblique firing becomes impossible, and, in this case, a column will oppose as much fire to the cavalry which charges it, as a thin battalion; for the three first ranks of the column are as able to fire as those of the battalion.

I know I shall be told of the many instances of deployed infantry repulsing cavalry, but surely the latter must have wanted courage in those instances. All the officers of cavalry, who have seen service, declare unanimously, that, in general, their troops do not retreat, till after they have received the fire of the infantry, that is to say, when there is almost nothing more to fear. This conduct is unaccountable: it is doing too much, or too little. If the troopers, after receiving the fire, were to clap spurs to their horses, and give them the bridle, they would penetrate the ranks. In general, the fault is thrown upon the horses, which, it is said, will not advance when once they are seized with fear. Certainly they have no means of exculpating themselves: but the fact is, that these animals are very warlike, and, when their spirit is roused, rush undauntedly upon the bayonet, as we have often seen. Do not let us forget that the Roman knights, in a battle with the Samnites, I think, unbridled their horses, in order to pierce an infantry, till then impenetrable, and the trial succeeded.

Indeed, I do not think that horses in the ranks can be easily repelled. A single foot soldier will defend himself, with his musket and bayonet, in an open spot, against a single trooper; but it is not the same thing with a troop of regular cavalry, charging modern infantry.

How many square battalions, composed of brave and disciplined troops, might we not count, that have been broken and dispersed by cavalry, in spite of the most obstinate resistance? I will here cite but a few examples: A square formed by the 14th Saxon battalion, at Langensalz, was broken, and the men made prisoners by hussars, though they defended themselves most bravely. The Prussians, under the command of General Fouquet, at Landshut, likewise formed into square battalions, which were broken by the Austrian cavalry, and the men were all either cut to pieces, or made prisoners. So, in the year 1793, in the war of the revolution, the Austrian cavalry, near

I.

Quesnoy, crushed a square battalion of French, who waited the attack with the greatest intrepidity. All wars furnish similar examples. There are likewise examples to the contrary: I shall cite that of the Manteuffel regiment of Prussian infantry, which retreated from Neustadt to Neisse, exposed to the continual attacks of the whole cavalry of Laudon, without suffering itself to be once penetrated. But we ought carefully to examine whether the failure of these attacks may not be imputed to the cavalry. It is a question that ought to be decided; yet, though I am persuaded that cavalry must get the better of the infantry of our times, if local circumstances are not entirely in favour of the latter, I would not hazard a peremptory opinion; but I wish that experienced military men would publish the result of their experience and judgment on the subject, that a final decision may be pronounced on this military problem.

Indeed, it must be said in favour of the infantry, that there is no ground where it cannot act, while the cavalry find few places convenient. To this remark, I further add, that infantry, scattered as *Tirailleurs*, acts with much more effect, in intersected countries, than infantry in close ranks.

It seems to me demonstrated, that the augmentation of the cavalry in our days, when compared with that of the infantry, is owing to a consciousness of the former being superior in strength to the latter; for, we may reconnoitre the enemy in countries entirely open, of which there are few, with infantry, as well as with cavalry; and in close and intersected countries, infantry is much more useful; nay, most generally it is only infantry that can be used in such situations: nor is cavalry equal to longer marches. I conclude, therefore, that it would not have been so much augmented, but for a confused notion of its superiority in combats, where the ground is not unfavourable to it; a superiority which must be attributed only to the mode of arming the infantry.

I am of opinion that this augmentation was absolutely necessary; for, had it been adopted by one power and not by the others, the latter would have found themselves badly off in their contests with the former. From the manner of disposing and arming modern infantry, it is impossible to bring it forward, without having cavalry at hand to support and cover it, in case of a retreat; especially, if the enemy has cavalry, and the nature of the ground does not render it quite useless. The celebrated example of the English and Hanoverian infantry, which, at the battle of Minden, overthrew the French cavalry, is no proof against what I advance, for it would be necessary to ascertain whether that cavalry had not committed capital faults.

The result of what has been said is that, in general, the cavalry ought to be placed behind the infantry, unless particular circumstances modify this rule: for, if the enemy has his infantry supported by cavalry, and yours has not the same support, it is evident, that if your infantry is put in disorder, it must be cut to pieces, if the enemy sends his cavalry in pursuit of it. I am aware that a troop of horse loses its greatest advantage, which consists in its rapidity, when it is engaged with a body of infantry; if the affair, however, takes place in a ground favourable to the former, it knows no curb, and its effect is incalculable. It has this advantage when posted on level spots between woods and eminences which the infantry have been made to occupy. Here the cavalry, protected by the fire of the infantry, resembles a curtain between two bastions. It is likewise proper to place cavalry in a camp, in such a manner that it may be entirely behind the infantry, and not upon the wings, unless on each flank it has infantry whose fire, crossing in its front, may cover and protect it; for, the horse stand in need of the foot in a camp, because not being able to form in order of battle quickly, they are in danger of being cut to pieces if taken by surprise. When once mounted, the cavalry can defend themselves in an open country better than infantry; but, before they can get ready, they have a hundred things to do; whereas the foot soldier takes up his musket and is prepared for action.

It is the nature of the ground, then, that shows the points where cavalry ought to be placed in the day of battle. Intermixed with small corps of infantry it loses its spring, and consequently its vigour; but preserves these advantages, when posted in a second line behind the foot. It emboldens the latter, which finds itself supported, and covers its retreat by a very slight movement; for it is very easy for the infantry, to retire through the intervals of the squadrons, and indeed *Tirailleurs* may effect this manœuvre, running at full speed. In marches in column, the mixing the two kinds of troops is very proper, especially if the ground is diversified, as it almost always is. In a plain, it is the part of the cavalry to meet the enemy; in woods and mountains it is the part of the in-

fantry; in a mixed country, it belongs sometimes to the one and sometimes to the other,
according to the variations of the ground; and
it is on this account, that in such a country
they ought to be intermixed, when marching
in column, so as to be able to advance one or
the other, according to circumstances. Were the
whole cavalry to march first, and the infantry
follow, the latter would require too much time
to advance in front, if it were wanted. Thus,
in perpendicular marches, that is, where the
heads of the columns are turned towards the
enemy, these two kinds of troops should be so
disposed as to be ready alternately as they may
be wanted by the general. This is not necessary in parallel marches by the flanks; for, in
this case, as the columns march open, not
having to deploy, but merely to make a quarter wheel to form in order of battle, it is very
easy for a column of cavalry to change places
with a column of infantry, according as one or
the other is wanted to face the enemy first.

I shall now resume the consideration of the
oblique order of battle, from which I was
diverted by the chain of ideas. I was examining the manner to be pursued, to throw a
wing back into a line, to prevent its being
turned. Let us suppose then an oblique line

as C D (*fig.* 45.) with a hook *d e,* formed as a
security against being taken in flank at *d,* and
to have a line ready to repulse all the attempts
which the enemy A B might make with his
left on *d,* which he outflanks.   This is the
first modification of this kind of position, which
presents itself for our examination.   After
this line, which gives the appearance of a
hook, or rather of a gibbet, has driven off all
in the way of its march, it turns till it arrives
on the prolongation of the oblique front C D,
and then takes the enemy in flank.   This
movement would be equally necessary and
much easier, supposing it to meet with no opposition.   Epaminondas had something similar
at Mantinea; but as he broke the front of the
enemy with his flank, a thing now impracticable, he probably adopted this measure with
the view of preventing the enemy's wing from
enclosing his columns by wheeling.   Cæsar's
oblique line at Pharsalia was something different, and was intended only to defend itself.

At the battle of Lissa, some battalions of
grenadiers were placed at the extremity of the
right wing of the cavalry : they overthrew the
troops of Wurtemberg, and rendered other services.   But the oblique order of battle has
this defect, that it offers to the enemy a flank

which may be enfiladed by his cannon.   This would be the case with *c d*, as well as with *d e*, if the line A B outflanked and turned the oblique front *c d*; if it outflanked it greatly, it would surround *d e* by a single wheeling to the right. However, this precaution of having a hooked flank is excellent when you attack the enemy with an oblique close front, and in case you should not have completely come up on his flank. I will say more; it is necessary.   But still, it would not be prudent to advise such an attack against a skilful enemy; for he may render it useless by the means above mention- ed, as well as if he had only to act against an oblique line without the hooked part.

If the flank of the attacking wing were covered by a square battalion, as in figure 46, the cannon of the enemy would, without a possibility of preventing it, enfilade the two sides *e f* and *d g*, which is a very great disad- vantage.   Besides, a square battalion marches much more slowly than troops drawn up in a line.

An oblong, no doubt, marches quicker, but the long sides of this figure are no greater de- fence than a line; and the others, being short, may be easily turned.

The defence of an oblong, then, is weaker

than that of a perfect square. What is called the *Crèmaillière* (*fig.* 47.) which for some years has been in use in the Prussian armies, is too complicated to move easily, and is exposed to be enfiladed by the cannon of the enemy on too many sides. Yet, it is sometimes thrown into the form seen in figure 48, which is still more complicated than the other; and here we may ask why the thing is not perfected into the configuration of a rose, as the Chinese do it, if Captain Tielke is to be believed [*]. The end aimed at by such a number of flanks as are obtained from the *Crèmaillière*, is to dispense with the oblique firing, because the direct fire of those flanks sweep the front of one another; and it is clear, that the soldier who has an enemy before him, will not fire obliquely. The first thing he does with his shot is to take care of himself, before he thinks of his neighbour. But the cross fire, which results from this angular formation, may produce no great effect. Further, it is difficult for several platoons to fire together, at the point where the meeting of two fronts form a re-entering angle;

---

[*] It is hardly necessary to observe, that the author, who in this paragraph criticises the *crèmaillière*, is here jesting.

COMMENTATOR,

for, they would fire in one another's faces if that angle be less than 90 degrees; and if it be obtuse, the cross fire would tell very little.

Some will say, that in order to strengthen a square battalion against cavalry, it may be formed thus: let the third rank be separated from the other two, and let it form an inner square of itself. (*fig.* 49.)

This done, if the enemy's cavalry should penetrate one of the angles of the outer square, the inner square (*fig.* 50.) would form, by wheeling to the right and to the left, as the dotted lines show, a re-entering angle, which by a cross fire would drive the enemy back. But it is not at the angles, where the cannon are placed, that a well-conducted cavalry would attack a square of infantry; and if the sides are assailed, the third rank, though separated, as we have said, from the two others, and though a fire in reserve, would be of little avail.

It has also been proposed by many, to make retreats in several square battalions; but the squares are required to be small, composed at most of two or three battalions; and they are to be so disposed, during the march, as to flank each other. (*fig.* 51. 1, 2, 3.) The battalion 3 clears the front of the battalion 1, which renders the same service to the rear of 3, and to the front of 2, while 2, in its turn, protects at once the rear both of 1 and 3. This is all very fine on the parade; but, pursued and assailed by an adventurous cavalry, it would be difficult for these different squares in the field to keep, in respect to one another, a position so laid down by line and rule; and it would be much to be feared, that they would treat one another with some balls, in their cross fires on their fronts.

Men experienced in the art of war have, nevertheless, preferred for retreats of the infantry, square battalions to the chequered retreats in line, (*fig.* 52.) This manner is so well known, that I shall not describe it; I shall only observe, that, in the execution of it, the distances have never been preserved; not even in the manœuvres of the Prussian infantry: especially, when it was necessary for one of the wings to bear aside by wheeling during the retreat (*fig.* 53.) in order to prevent the pursuit of the enemy, by alarming him for his flanks. In other respects, retreats of the latter kind being eccentric, are consistent with good principles.

When, after engaging with musketry, you want to retreat, it would be in vain to pretend to do it perfectly in order. In such cases, the troops will always fly in confusion, for otherwise, there would be no occasion for their quitting the field of battle. It is on this account, that it becomes necessary to have a line of cavalry behind the infantry, to support it; and, then, it is not so prejudicial as imagined, to fly in haste among the cavalry: on the contrary, the men are sooner out of their dangerous situation. It is only necessary, that this dispersed infantry should immediately form again in the most convenient place; in a wood, on a height; and if it returns directly to action, it displays a better founded courage than if it had retreated inch by inch, and lost a great number of men: for, in the one case, the courage is useful, and in the other it tends to nothing. If in an open situation you have no cavalry to support you, then you must, no doubt, keep your troops together, or they will be cut to pieces.

If there has been nothing but cannonading, which among the moderns is also called an action, an orderly retreat becomes easy, even if a slaughtering fire of musketry should afterwards take place in the course of it; and I shall here observe, that whenever a regular retreat can be effected, the best and easiest thing to do, is to face to the right about with the whole line, and thus march retrogressively. In this manner you get sooner rid of the enemy's fire, than you could in a chequered retreat, and the order is more easily kept, which deserves to be considered.

I know no situation more justly exciting pity, than that of a square battalion surrounded by *Tirailleurs*, (*fig. 54.*) All the shots of the latter are concentric, consequently of the greatest effect; all those of the square eccentric, and nearly of no effect. The ranks of this unfortunate square would be very soon thinned by well directed fires, every shot from which would do execution, and a battalion in this position could not be saved from ruin.

The most celebrated of the modifications given to the oblique order of battle, is the oblique attack in *échelons*, invented by Frederic II. It would be unpardonable in me to pass it over in silence. But let me first observe, that Frederic gained no victory by the rules of the *échelon*. He conquered at Leuthen, not because he attacked the Austrians in *échelons*, but because he took them in flank*. At Zorndorff it was

* At the battle of Lissa, the King of Prussia, with his picked troops, attacked the left wing of the Austrians, which he took in the rear, and defeated while he availed himself of a ridge of heights opposite to their right and

attempted to make the left wing effect this manœuvre, but it did not succeed, and the eight battalions detached from the rear-guard were not properly disposed according to this form. Experience has not yet taught us the excellence of this mode of attack, and Capt. Rösch has shown that it is not maintainable on principle. He demonstrates that each *échelon* will receive a superior fire from the enemy ; for, if the *échelon* c d (*fig. 55.*) advances within musket shot of the line A B, it will suffer extremely on its flank c, and will therefore receive a greater fire than his own. The flank c, which will find itself exposed to an oblique fire, will insensibly describe an arch backwards, to face its adversary. The platoon of the line A B, which, in this case, pours so destructive a fire on the flank c of the *échelon c d*, will not be prevented by the second *echelon e f*, this being at too great a distance to fire ; nay, the two first platoons of the wing *f* would hardly venture to fire if the *échelons* were three hundred paces distant, for fear of hitting the flank c. Thus, the two opposite platoons of the line A B, would   shower

centre, to mislead them, to keep them in check, and to station there, in an excellent posture of defence, the rest of his army which he had weakened by detaching reinforcements to his right wing —COMMENTATOR.

their balls without the slightest hindrance, on the unfortunate flank *c*. If the *échelons* are only fifty or a hundred paces distant, these inconveniences will not take place; but then, on the other hand, the advantages expected from an attack in *échelons* will be lost. Those advantages are, that, by dividing your front, you expose only a part of it to be defeated, as you refuse the others; whereas, in an uninterrupted oblique line, disorder quickly propagates itself from one end to the other. Now, if the *échelons* follow one another only at a small distance, you no longer refuse any thing, and most of the line is at the same time exposed to the fire. It is for this reason, that they are placed at the distance of two or three hundred paces; and then we have seen the sad consequences to them; that is, their flank is exposed to a most destructive fire, as they advance. This kind of attack, therefore, is good only when you have to meet an enemy stronger than yourself, and if you have the superiority, there is no doubt, that to attack your adversary at once in front and on his flanks, is infinitely more efficacious and more advantageous.

Indeed, it is possible to turn the attack *by échelons* to better account, by having the first, and second *échelons* supported by a second line, and leaving the others weaker; this is, in my opinion, the only method to use *échelons* with advantage. But still, the second line would prove of little avail, were it composed of infantry; for, it would be useless, while the first line stood its ground; and if it were put to flight, it would probably carry the other with it. Indeed, when infantry is drawn up chequerwise, the first line, if beaten, may fly without throwing the second into disorder, on account of the great intervals between the divisions of each line; but, this is not the case with the long, thin phalanx of Frederic II.

Besides, I defy any one to mention a single instance of a second line of infantry renewing the fight, and taking the place of the first, where the latter was defeated; now, according to rule, the first *échelon* must necessarily be beaten, as it is exposed to a concentric fire; and if the battle be continued with steel, which, indeed, is not common, it is only requisite that a platoon of the line A B should, by wheeling, come on the flank of the *échelon* c d while the front is engaged, to overthrow that *échelon* before e f, three hundred paces off, or even the second line of the *échelon* c d, could come up to re-establish things: thus, the line A B will, in all probability, defeat the *échelons*, one after the other, and that the more easily, as they will be taken in flank the moment *c d* is obliged to fly.

More: if the line A B puts itself in motion by its right wing, and attacks the *échelons* intended to be refused, as these will be obliged to face the enemy, and to form in a line parallel to his, the consequence must be a battle in front, the fate of which depends on chance. Were the *échelons* to maintain themselves in their position, they would be taken in flank by a superior force, and victory would soon declare for the assailants. It is very easy to counteract the obliquity of a line of battle, and as for *échelons*, they are generally good for nothing.

I grant, however; that *échelons* have this advantage: that their flank is better secured from the enemy, were he to advance in order to attack them, than the flank of a continued oblique line would be; for *échelons*, by simply wheeling (*fig. 56.*) may keep their front in a direction constantly parallel to that of the enemy, and thus secure their flank, which cannot be attained in a continued oblique line, but when the flank is completely upon that of the enemy, as has been explained.

I shall further observe, that it is impossible
to come upon the enemy's flank by the diagonal step, executed during the march, if,
before you put yourself in motion, you do
not considerably out-flank his wing; for, if he
marches flankwise on the prolongation of his
front, he obstructs your design: he will run
over, in the same time, a greater space than
you can do with your diagonal step; because he moves on a direct line, and you for-
wards and obliquely together, which shortens
your step; and because he marches on one of
the smaller sides of a right-angled triangle,
while you move on the hypothenuse, which is
longer. It is impossible, then, to succeed in
coming up by the diagonal step on the flank of
the enemy, while you march in order of battle,
if he knows how to act*.

* I admit, with the author, the defects of the oblique order
of battle, in a single continued line; and even grant, that *échelons* are not faultless; but, it is possible that an army, though
not really placed in an oblique direction to the front of the
enemy, may nevertheless, either from the advantage of the
ground, or the skilfulness of its movements, place itself in a
situation to attack him at one or more points; and at the same
time to refuse the parts of its line, which are not intended
to be engaged. Such a disposition seems to have a great
analogy with the oblique order of battle in general; at
least, as to the effect it produces, and the end proposed

Finally, cannon, fired concentrically, are able, no doubt, to give infinite effect to the force of an attack: but this kind of support is not peculiar to *échelons*, and may be adapted to any order of battle. In the attack by *échelons* (*fig. 55.*) batteries should be placed in front of *e f*, in order to enfilade the part of the line A B, which would endeavour to fall back into the line *i k*, if it were taken in flank by *c d*.

The most, and, indeed, the only useful step to be taken to support an attack, is to place a line of cavalry behind a line of infantry; it covers and secures the retreat of the latter, if unsuccessful; and if the infantry of the enemy be beaten, it throws it into complete disorder.

Cavalry, placed immediately behind infantry, supports an attack still better, than if there

in forming it. I shall here remark, in respect to this disposition, that, besides its being that which an army is commonly under the necessity of taking in war, as the ground is rarely sufficiently level and open to admit of a regular obliquity, it affords likewise very great advantages in the opportunity it gives of profiting by local circumstances, either to deceive the enemy, or to secure the parts of the line which have been weakened, for the purpose of reinforcing those to be sent against the points of attack.—COMMENTATOR.

were a second line of infantry between them, for, in the first case, there is nothing to prevent its advancing to the assistance of the flying troops, and covering them; thence it follows, that there should be but two lines, one of infantry, and one of cavalry; and this the more, that two lines of infantry being of no avail, except the second is out of cannon reach, the latter should rather be considered as a reserve, than a second line in action. The idea of giving strength to the *échelons* by double lines of infantry is consequently an error.

But once more let me say, that, to conquer an enemy, the most efficacious measure is, to attack him in flank, contenting yourself to amuse him in front by detached corps; and the columns ought to be prepared for this chief attack, out of sight of the enemy; that is to say, *strategically,* just as an admiral works at a great distance, to gain the wind of the enemy's fleet. No manœuvre within cannon reach could have a successful result, allowing the enemy to be dextrous. And in this point of view, I think the battles of Crévelt, Friberg, and Leuthen, real masterpieces of military skill.

Pursuing the idea of this division into different corps, I am led to examine an attack in

three corps, extolled by Folard, and considered
by Lloyd as the best. Let me first observe,
that a greater division of the forces is rarely
necessary. Two corps are employed to manœuvre on the flanks of the enemy, the third
is directed against his front. This is the conduct to be pursued by an army when stronger
than the enemy; when, on the contrary, it is
weaker (the case admitted by Folard and Lloyd)
it is to place its three corps opposite the two
wings and centre of the enemy; so that,
even in this case, three are best.

The Chevalier Folard proposes, with his
centre to break the enemy's centre, and with
his wings, to prevent those of the enemy from
taking in flank the corps which attacks his
centre. I will examine Folard's opinion in
the following chapter; Lloyd's has many new
ideas. Proposing two lines, he places the
cavalry in the second, and forms his infantry'
four ranks deep. But what astonishes, in a
man so profound in the art of war, are the
lances he advises to be fixed to the firelocks,
at three hundred paces from the enemy, in
order to charge his infantry with this new wea-
pon. How could such an invention resist a vigorous fire? General Lloyd particularly destines
the wings to cover the centre; they are to be

chiefly composed of cavalry, intermixed with small columns of infantry. According to the opinion of this great soldier, it is by this ar- rangement that the infantry will best resist the enemy's cavalry; and he asserts, that an infan- try of three men deep, is not able to stand against his of four with their lances.

I have, in this chapter, given my opinion on the objects of Tactics. It was not my inten tion to treat of them very circumstantially, as I wished not to make too large a volume. I trust I shall be excused for the length of this chapter, considering the many points I have examined. I do not pretend to regulate the extent of my chapters, but I follow, without system, the chain of my ideas. If I am charged with ad- vancing a paradox, in declaring a preference for *Tirailleurs* over infantry of the line, I shall support my sentiment, by citing a book entitled *Considerations on the Art of War*, the author of which, who served in the seven years war, expresses himself in the following manner, on the engagements of infantry, in that celebrated war:

" It was never possible to depend upon a regu-
" lar fire by platoons, divisions, or battalions;
" it was with difficulty the troops fired with
" order even once; they then immediately began
" an irregular fire, which was continued accord

" ing to the dexterity of each man in handling
" his firelock. The hind ranks frequently ad-
" vanced and fired over the shoulders of those
" in front; the first rank never knelt; the troops
" formed into five or six ranks, sometimes more,
" sometimes less; the line became a mass, with-
" out order, and the general and officers waited
" patiently for the result, to know whether the
" battle were gained or lost. And it is remark-
" able, that it was the Prussian infantry that
" presented this picture." If, then, in battle,
the fire of infantry of the line so soon degene-
rates into the irregular fire of *Tirailleurs*, I see
no reason why I should not say that soldiers
ought to be trained to this exercise; or why I
should not prefer a corps of organized *Tirailleurs*,
to one compelled to become such by circumstances.

I shall next examine Folard's system.

CHAP. XII.

Of the Chevalier Folard's Column.

THE weakness of the flanks of an infantry drawn up in a slender order could not fail to lead a mind so penetrating as that of the Chevalier Folard, to inquire into the means of remedying the evil.   He judged that the front of an infantry formed three or four men deep was still too weak.   On this reflection, the deep order appeared to him the best calculated to defeat troops disposed according to such false principles; and from this arose the idea of his column, which he presents to us as a mass of infantry, the flanks of which are longer than the front; supposing, however, that the column consists of more than one battalion.   Each battalion forms nearly a square. (*fig. 57.*)   The most considerable columns of Folard are in three divisions; as *a, b, c*.   The utility of this disposition is, that the division *b*, for example, having it in its power to move aside, as towards the point *d*, a cross fire takes place, the fires of d being directed along two sides of c and of $a*$.

* The movement of the division h is particularly applicable in the case where a column is attacked and must defend itself. In the offensive Folard, in general, makes little account of fire ; the principal effect he proposes from his column being to break by shock the front of a deployed nfantry, against which it is to march rapidly.

The author's description of Folard's column appearing to me insufficient to give an idea of it, I think it necessary to make some addition to what he has said.

Folard supposes the front of his column generally composed of thirty files. I say generally ; for he admits of differences according to the nature of the ground on which it acts. In an open spot, for example, the number of files may vary, in his opinioh, from twenty to thirty, and on covered ground, and where there are defiles, from sixteen to twenty-four. He thinks a front of greater or less extent defective. Let us suppose it to be thirty files : he divides this front into two equal parts, which he denominates *manches*, and which are consequently fifteen files each. He distinguishes them by the name of right *manche* and left *manche*, and divides the front of each *manche* into three sub-divisions of five files each, which he calls *divisions of the right* and *divisions of the left*, according to the *manche* to which they belong. The first and last divisions of the front of the column are named *divisions of the wings ;* the two following, *the first division of the right*, and the *first division of the left*; and lastly, the two other divisions, which are in the centre, he calls *the third division of the right* and *the third division of the left*. According to Folard's system, each column may be from one to six battalions

The following is the manœuvre which Folard makes his columns execute:

After piercing the enemy's line which it has broken in front, the column separates lengthwise in two halves, one facing to the right, and the other to the left, such as *a* and *b* (*fig.* 58.) and they then attack in flank the separated parts of that line.

It is proper to observe here, that by this move-

strong, every battalion being supposed to consist of five hundred men, exclusive of a company of grenadiers, the officers and serjeants, whom he distributes in the following manner: Six captains six lieutenants, and six serjeants, at the head of each of the grand divisions of the column, and the rest behind, or on the flanks Of the five hundred men, four hundred are armed with muskets, and the other hundred with a kind of spear, twelve feet and a half long, (including both staff and steel,) which he calls *pertuisanne.* He does not place the grenadiers in the column, but in the rear or on the flanks of the last division: and he designs them either to pursue the enemy, after his front is broken, or to fall upon the flanks and rear of the battalions and squadrons which should continue resisting; or for any other use, which, according to circumstances, the general of the army may think proper to make of them. Lastly, he considers them as serving for a support and reserve to each column.

I have entered into these details respecting Folard's column, because it has a great analogy with the columns of attack of the present times, and because these were taken from it —COMMENTATOR.

ment, Folard acknowledges the advantage of enclosing the enemy by outstretching his front; as
in this manœuvre the Chevalier's column, the
flanks of which are longer than its front, always
out-flanks the divided parts of the enemy's infantry which he supposes to be only four men
deep, and which, of course, can, on its flanks,
present only a front of four men.

The movement made by the two halves of the
column, the one facing to the right, the other to
the left, after breaking the enemy's line, is not
so simple, and consequently not so good, as that
executed by the division *b*. (*fig. 57*.) It is impossible, that this separation of the column can
be effected with order in a battle.

An objection has been made to this column,
which does not appear to me well founded. It
is said, that a number of men drawn up behind
one another, does not move like a wedge driven
by a mallet, because each individual possessing
within him the spring of a spontaneous motion,
forms in a manner an independent whole. But
the objector does not recollect, that the foremost
ranks of a column are induced to advance by
powerful motives; for they are pushed on by
those which follow, and which from impulsion
would crush them, if they attempted to stop.
When a crowd pushes to a door suddenly open-

ed,' those who are foremost are compelled to follow the direction of the torrent. It is for the same reason, that infantry in deep order overturns another in slender order. And if they come to bayonets, the latter must be overpowered: for, three ranks must yield to the efforts of twenty. But, in fact, they will not even come to bayonets; for, the pressure alone of the profound mass moving forward will be enough to throw the three ranks out of their position.

It has been said, and this cannot be denied, that cannon, firing grape shot in different directions on a close column, would inevitably destroy it. But I must observe, first, that as a column advances more rapidly than a deployed line, the effect of cannon against it is consequently diminished. Besides, a cross fire of artillery would be as destructive to battalions deployed. The battle of Torgau, and other examples, prove this assertion An extended front is like a great ring, which it is more easy to aim at and to hit, than the narrow front of a column; even allowing the balls to do more havock in the latter, than in the former. How many balls pass through the intervals of a column, which would tell in a line of infantry. It is not considered, that there is no means of attacking, in front, batteries which fire grape shot; in these cases you must take your fate, whatever order you may have adopted. It is only cavalry that can undertake any thing against those batteries with any hope of success: because, going with more speed, they are not so long exposed to the fire. Every thing, in the modern system of war, appears to confirm the importance of cavalry.

On the other hand, if there is an effectual means for the foot to resist mounted troops, it is unquestionably that of forming into a column; and to the present mode of arming the infantry, is attributable the power still possessed by the cavalry, of penetrating columns. I am persuaded that infantry armed with pikes, supported by shorter weapons, as advised by the Chevalier Folard, would make an unshaken resistance against cavalry. But it will be said, that if the enemy has cannon at hand, the column will be injured by their fire, and then crushed by his cavalry, if it has none itself. True; and it is another proof, that since the invention of fire-arms, every thing conspires to give importance to cavalry, when the ground is favourable. From these considerations, then, it appears that the criticisms of the modern Tacticians on the system of Folard are raised on a foundation

of sand; a côlumn fails only againtt *Tirailleurs,*
But, as this is the case likewise with infantry
deployed, it proves nothing against the deep
order.

I have stated the advantages which result
from the formation into columns, for attacking
a deployed infantry.

Indeed, these advantages will be rendered ineffectual, if, instead of resisting in front the
attack of the column, the enemy fell upon
its flanks. It is against columns that *Tirailleurs* have particularly the most terrible effect.
As they then fire upon a mass of men, almost
all their shots will tell. Let them yield the
ground to the column as it advances; let them
keep at a proper distance from it; let them fly
about it, trimming it well with an irregular fire,
and its destruction is certain. If there are several collateral columns, as must necessarily be
the case in a large army, and if they are supported by cavalry in a second line, it is then
no doubt impracticable to penetrate between
those different columns, in order to take them in
rear and flank. But still, the rule of not waiting for the shock, must be attended to. You
must move aside, you must menace the flanks
of the formidable colossus that advances; and
your movements must be skilfully prepared

while you are yet at a distance from it: and
as it is always easy to avoid a direct attack by
moving aside, and then coming up again upon
the flanks and rear of the enemy, the consequence is that the advantage of the columns,
which can only be felt in fighting in front, is
entirely lost.   In flat countries, these movements are covered by the cavalry, and should
those attached to the column advance,   the
result would be an action between the mounted
troops, where the advantage is equal on both
sides.

The excellence of the plan proposed by the
sagacious Folard, of placing cavalry in a second
line behind the infantry, has at length been ac-
knowledged.   However, it is proper to observe,
that the method he advises of ranging the cavalry in deep order, proves that he was not
thoroughly acquainted with mounted troops.
He likewise proposes to fill the spaces between
the squadrons with infantry ; but, as he requires
it to be light infantry, the disadvantage of this
disposition becomes less : because this kind of
infantry may be trained to run with the cavalry.
It is undoubtedly of great use to support cavalry by infantry.   The French, since the Revolution, have always had *Tirailleurs* attached
to their horse.   It was a custom with the an-

cient Germans : and, indeed, no party of cavalry should be detached, without the attendance of light infantry.

We see that Folard is a great advocate for the *petite guerre*, or war of parties, by the tribute of admiration which he pays to Sertorius. He strongly advises the system of cutting off provisions, avoiding battles, acting on the rear and flank of the enemy, alarming him by night, &c. He would have given his opinion on this practice still more decidedly, had he lived to see the new system arrive at greater maturity.

I shall conclude this Chapter with saying; that *Tirailleurs* are the only troops more conformable than columns to the modern system of war; the best way seems to be, to employ sometimes the one and sometimes the other.

CHAP. XIII.

Difference between the Tactics of the Ancients and those of the Moderns.

WE observed in the beginning of this part, how much the Strategics of the ancients differ- ed from those of the moderns, from the wants of their armies being very moderate compared to those of ours. We are now going to examine, though briefly, the difference between their Tactics and those which we have adopted.

The principal difference is, that theirs related entirely to engagements with steel weapons, and ours are adapted to those with fire-arms. Hence arose among them, the superiority derived from corporeal advantages, and the necessity of the deep order, to overthrow the enemy by its weight, and to bring on new troops against him as the foremost fell; hence, too, the narrow front of their armies; for they had nothing to fear from cross fires, as we may admit, without hesitation, that the effect of missive weapons was of far less im-

N

portance among the ancients than among the moderns.

Formerly, then, they might endeavour to break the centre, which generally decided the fate of a battle, though the attempt did not succeed at Cannæ. No fear was entertained of an enemy's movements on the sides of an army. In our days such movements would endanger its supplies, but alarms of this nature were unknown to the ancients.

It was rather the bravery of the soldier than the ability of the general that formerly decided the victory. Cæsar, in respect to his battles, deserves admiration less for the skill of his positions and movements than for the manner with which he inspired his warriors with intrepidity and ardour.

Battles in those days were much more bloody than now, because, from the nature of the deep order, a greater number of men were engaged at once, and because the moral energy of the ancient soldiers rendered their struggles long and terrible. At present the whole affair is over, when the one army has turned the flank of the other* Independently of this,

---

* Generally speaking, the author is very right; and I am far from denying that an army turned on its flanks is not in a critical position; yet it is not impossible that, by

there are fewer men slain by fire than by steel: bodily conflicts are rare and soon over among modern troops, from the manner in which they are armed: the most bloody, and that only on one side, are those in which cavalry cut down broken infantry. Between two corps of foot the shape of the musket and bayonet renders them more ridiculous than dangerous.

Among the ancients, when once the armies came to action, a retreat in order was almost impossible. Let me not be told of *the retreat of the ten thousand*, and some others, which entirely owe to very peculiar circumstances their being considered as exceptions. Nor had the ancients numerous cavalry to protect their flying infantry: besides, their cavalry could do nothing against the infantry, armed as the latter were.

Infantry, as they fled, were cut to pieces by soldiers who pursued with equal velocity; the slaughter must have been terrible. It was impossible to think of taking a position, or of facing any more, for they had no batteries under

skilful and rapid movements, it may recover itself: but then, it must understand manœuvring well, and its skill in Tactics must exceed the narrow bounds in which, according to the author, it is circumscribed by the modern system of war.—COMMENTATOR.

shelter of which they might rally, and form
again; the enraged enemy left no time for such
a manœuvre.   The armies of the ancients had
not all the generosity of those of Lacedemon,
which, content with victory, did not pursue the
enemy.

If the system of Folard's column were generally adopted, battles would become extremely
bloody.   The imagination recoils at the idea of
the shock of two masses of men ranged in deep
order, and which, being broken soon after, would
present the view of a multitude killing one another without order or principle: for, the infantry of modern times, from the mode of arming and accoutring them, do not form a body as
compact as that of the ancients, which was
united by bucklers.   Thus, in the deep as well
as the slender order, an engagement of infantry
must, in a little time, become a fight in the
manner of *Tirailleurs*;   so much the more,
as in the present mode we are not to think
of being able to avoid fighting, when near the
enemy.

The battle of Cannæ proves that it was also
useful, among the ancients, to outflank the enemy; and Tacitus, in the 75th chapter of the
3d book of his Annals, has a passage which

proves to demonstration the excellence of offensive operations tending to surround the enemy even in the ancient system of war :

" Tacfarinas," says the historian, " generally
" divided his men into small parties, which
" made it easy for him to open when attacked,
" and at the same time to charge our troops in
" the rear. Upon which the Romans formed
" three columns, one of which was commanded
" by Cornelius Scipio, the Proconsul's Lieute-
" nant, who advanced with his division to the
" quarter where Tacfarinas, ravaging the coun-
" try near Leptis, fled for shelter to the Gara-
" mantes. In another quarter, where Cirta lay
" exposed to the Barbarians, the younger
" Blæsus, the Proconsul's son, commanded a
" second division. In the intermediate part of
" the country, the commander in chief march-
" ed at the head of a chosen body of troops.
" At all convenient places he threw up en-
" trenchments, and appointed garrisons, to
" check the enemy and frustrate his projects;
" for the Barbarians, wherever they appeared,
" found themselves counteracted on every side :
" the Romans were at hand in front, in flank,
" and in the rear; so that a great many were
" either put to the sword or taken prisoners.
" After this, Blæsus, the Proconsul, subdivided

" the three divisions into smaller parties, under
" the command of Centurions of approved va-
" lour and experience.   Nor was the campaign
" closed, as usual, at the end of the summer.
" Instead of retiring to winter-quarters in the
" old provinces, Blæsus kept the field: he in-
" creased the number of his posts and garrisons,
" and sent out detachments, acquainted with
" the country, to pursue Tacfarinas, who, no
" longer able to oppose him, shifted his huts
" and wandered from place to place. At length
" his brother was taken prisoner, and Blæsus
" thought proper to close the campaign."

This instance in the history of wars proves
the excellence, even among the ancients, of
acting upon the flanks of the enemy, with the
view of surrounding him.   However, the armies
of the ancients, both as to numbers and con-
sumption, being able to live any where without
great magazines, had not to apprehend diver
sions on their rear nearly so much as the moderns, who are completely ruined by them.
The necessity of moral and physical interior
strength among the ancients, and the necessity
of great extension among the moderns, appear
to me the essential causes of all the difference
in the two Tactics.

CHAP. XIV.

Result of the foregoing Chapters.

THE following principles, relative to the modern art of war, arise from the preceding inquiries.

It is necessary to have magazines, and fortresses to secure them.

It is no less necessary to have a range of fortresses, on the same line, to serve as a base.

In order to undertake safely an offensive operation against the enemy, it is requisite that the two fortresses at the extremities of that line, be situated at such a distance from each other, that the two lines of operation, proceeding from them, shall, when they meet at the *object* of the operation, form an angle of 90 degrees, at least.

The progress of the enemy is better impeded by taking a situation on one side, than by placing yourself in front of him.

You must never suffer an offensive operation, and be contented with defending yourself; you

must assume the offensive, and make diver·
sions on the flanks and rear of the enemy.

You must, as soon as possible, relinquish parallel positions, and parallel defensive marches,
to follow the mode of diversions, of which we
have been speaking.

The supplies of the enemy's army, rather
than the army itself, should be the aim of
the operations.

It is easy to deduce, from these various strategic rules, what ought not to be done; as all
that is contrary to them, is bad. Thus, it is a
fault not to secure a sufficient base, to operate
on a single line, in an acute angle, &c.

As every offensive operation ought to be concentric, every retreat ought to be eccentric.

All these rules of Strategics are applicable
to Tactics, substituting the line of battle for
the base, and lines of march and of fire for
lines of operation.

It is always possible to avoid a battle, by
not suffering the enemy to approach too near.

You should never wait to be attacked in your
position, but always put yourself in motion to
attack, even though your position be one,
from which you cannot be driven by force:
as there is no position but what may be turned.

You ought only to amuse and check the

enemy's front; your serious attack should be directed on his flanks.

It is requisite to enclose the enemy; that is to say, to have a larger front than he has.

He is enclosed when you are on his flanks, even should you be much inferior in number.

There is more effect in fighting as *Tirail-leurs*, than in close ranks; and, besides, it is much easier to throw the latter into disorder.

As troops extend more as *Tirailleurs* than in the other mode, it is, of course, easier for the former to come upon the flanks of the enemy.

Infantry should constantly be supported by cavalry. The best means of doing this, is to place the latter in a second line behind the former.

A column is the best form of defence against cavalry. You should, therefore, act as *Tirail-leurs*, or form in column.

But experience shows that cavalry, when determined, vanquishes even columns of infantry, which proceeds from the mode of arming the latter.

Consequently, infantry must never be left without cavalry to support it, even in countries which appear impracticable for horses.

Retreats after battle should take place eccen·

trically, promptly, and covered by cavalry; thus protected, retreats may be made in disorder.

After losing a battle, you should immediately think of fresh offensive operations. Not to be really beaten, you have only to believe that you are not so In this case, it is proper to begin the *petite guerre*, to avoid battles, and to be content with manœuvring.

In the next Part, we shall examine the consequences of the Modern System of War, which itself arises from the necessity of a base of operation.

END OF THE FIRST PART.

THE

SPIRIT

MODERN SYSTEM OF WAR.

PART THE SECOND.

CONSEQUENCE OF THE PRINCIPLE WHICH
REQUIRES A MILITARY BASE PREVIOUS TO
ANY OPERATION.

CHAP. I.

*Masses, that is to say, the greatest Number of
Fighting Men, and the greatest Quantity of the
Materials requisite in War, must sooner or later
among the Moderns give Success; and not, as
among the Ancients, the Superiority of Discipline
and of Courage.*

THE ascendant possessed by the greater number of troops over the smaller, is, in the modern
system of war, the inevitable consequence of
the necessity of not suffering your wings to be

outflanked, and of the advantage that results
from outflanking those of the enemy. If you
have more men than your adversary, and know
how to make a proper use of that superiority,
you will render all the bravery and skill of his
troops of no avail; for all that the best soldiers
can do is to conquer the enemies they have before them: they may overthrow them, no doubt,
but while they are gaining ground in front, they
will be attacked by others on their flanks; and
we have already seen how dangerous this movement is to him on whom it is made.

But with an inferior number, the more the
victors advance, the more are they exposed to
be surrounded and separated from their magazines. It is impossible but that they must be
incessantly occupied with the means of preserving a communication on which their existence
depends. If that be threatened, they must not
only suspend their progress, but must fall back,
and, instead of pursuing a defeated enemy,
make haste to effect their retreat.

Let it not be said, then, that with 30,000
men, perfectly brave and disciplined, one may
compel an enemy thrice as numerous to quit
the field. It may be so, if the larger army be
badly conducted, but in no other case whatever. The bravest and best soldiers must yield,

when you can send detachments to attack them in flank and overpower them with numbers.

Nor let it be thought a good answer to this, to say, that, in such cases, the smaller army too may send detachments to protect its threatened flanks; for, it will not gain ground by this movement, as it is of a defensive nature: nay, it is a sure means of making the army already the weaker still more so, and of causing it to lose the advantage it might derive from concentrated forces. It appears certain, then, that in a very short time the larger army will manœuvre on the flanks and rear of the smaller one, without being obliged to offer battle. I have demonstrated this proposition in the preceding part of my work.

Thus, among the moderns, victory is decided by number, and not by courage and skill in Tactics. But that number must be conducted with ability; for, when the fronts of armies meet in battle, the more disciplined will, no doubt, put the less to flight*.

---

* As, on the avowal of the author himself, the ascendant of the greater number over the smaller, of which no one doubts, all things else being equal, is still subordinate to knowledge and skill, he was wrong in saying before, and in so positive a manner: " That the greatest number of fight-

By the word *masses* I mean, as well the materials requisite for the support of war, as the number of soldiers. These must be maintained by those, which is of course the most important consideration. The quantity of provisions, of clothes, of arms, and ammunition, decides the victory, as much as the number of men. It is the multitude of men and materials that, in the wars of our days, secures the triumphs of an army.

But as every thing can be soon obtained when one has money, the quantity of this commodity becomes, in turn, of great weight; for such is the charm of gold, that it will procure for you, even in your enemy's country, what you want in your own. The spirit of commerce supports the connection of nations in spite of the disunion produced by war. I will not even mention the means of obtaining your ends by bribery. It has already been said by Montecuculli, that three things were indispen-

---

" ing men, and the greatest quantity of materials requisite
" in war, must, sooner or later, among the moderns, give
" success, and not, as among the ancients, the superiority
" of discipline and of courage." Had he modified his proposition, as he does here, he would certainly have rendered it very just, and would not have fallen into an apparent contradiction with himself.—COMMENTATOR.

sable for carrying on war; *money, money, and money again.*

However, the materials necessary to the carrying on of war are so far independent of money, that it is of great advantage to have them at hand, and to be able to collect them sooner and in greater quantity than the enemy: for a State, which, having more money than warlike stores, should be obliged to have these brought from beyond the seas, or from any great distance, would not be so advantageously circumstanced as the adverse State, which, though poorer in money, were richer in materials. The latter, having it in its power to collect more readily a formidable mass, will crush the former under the weight of it. Of these two States, the one better provided with money, but less with warlike materials, will pay infinitely dearer for a smaller quantity of those materials, than the other whose situation is the reverse ; for, the price of things necessarily augments in proportion to the distance from which they are brought.

It is a decisive point, then, to be able to collect a greater mass of men and warlike stores in a shorter time than the enemy; but, it is not enough to have these masses, they must be organized in the most advantageous form.

In the preceding part of this work, I inquired into the nature of that form. The principle of the base shows that the materials required in war should be spread, just as the soldiers are deployed before a battle. In order that these masses of materials should be really useful, they ought to be placed on a line beside and not behind one another: it is a principle of Strategics. But these provisions not being safe unless lodged in strong places, the line we have been speaking of is of little use, if it be not formed by a range of contiguous fortresses.

If the masses were equal, and managed with the same ability on both sides, a thing impossible, the form of their developing would then decide between them; that is to say, the forces possessing the longer base would triumph over those opposed to them.

CHAP. II.

*One Consequence of the foregoing Exposition, is,
that small States, in future, will no more van-
quish great ones, but on the contrary will finally
become a Prey to them.*

THIS proposition can no longer be doubtful
after what has been previously stated; for, as
the number and quantity of the means of war
must sooner or later give success, it is evident
that small States can do nothing against great
ones, which are better provided with the means
of victory.   If, as among the ancients, courage
and discipline could dispute success with number and quantity, and counterbalance their effect, we might still see small States overturning
immense empires.   But we have demonstrated
that all moral energy, all warlike corporeal advantages, collected in a small number, must
necessarily fail against a great one.   This,
however, depends upon being able to make
use of that superiority conformably to the pre-

o

sent system of war : a condition that must not be forgotten. We have, no doubt, seen in our days the weak resist the strong, but it has been always owing to some fault of the latter. Besides, it must be recollected, that the art of war has but very recently been brought to its present state, and that consequently great States will in future know better how to avail themselves of their strength.

Great empires not only contain more men and more warlike materials than small States, but their frontiers likewise are necessarily more extensive : and the frontiers of the former generally enclose those of the latter. All the military resources of a great State are displayed, in the most advantageous manner, against a small one : and we may therefore easily judge, on which side victory will remain.

I speak only of neighbouring States ; for, according to the nature of things, we attack the nearest before we go to the furthest : but, those which lie between two must declare for or against the more powerful. If they are against it, this alters · the question ; for, several small States joined form a great one. In this case, however, the advantage is still on the side of that State in which the government has the greater means of enforcement in its hands, by the concentration of power in a single body politic; which is not the case in the coalition of several independent States.

CHAP. III.

Europe will one Day be divided into several great States.

IN Europe there are already several Powers whose forces are nearly equal. It is not possible for them to destroy the political existence of one another. But among these great States there are several small ones, which, sooner or later, will inevitably become the prey of the former, as soon as these can agree among themselves upon the division and appropriation of them.

Were there but one great State in Europe, and were the rest very inferior to it, the giant would devour the dwarfs, and there would arise a universal monarchy, which, indeed, would soon again divide into separate States, if the wisdom of the government did not keep pace with the extent of the body politic. But, as there are already several contiguous empires in Europe, we must consider them as forming an aristocracy, whose necessity of mutual communication forms a common centre.

This order of things is not favourable to individual ambition for fame; for, it is no longer possible to shine in history, as the conqueror of a great State, by the superiority of talents. Thus, we shall see verified what has been said: " that to him who has much, shall be given still " more: but he who has little, shall be stripped " even of that little."

Some States, however, and even those that will maintain their rank in Europe, are not equal in power; but they are defended by natural boundaries, which cannot be passed with impunity. This we shall show in the following chapter.

CHAP. IV.

*The Military Energies of States not being unlimit-
ed, it follows, from the Principle of the Base,
that they must diminish proportionally the further they Remove from their Source. They cannot act vigorously beyond certain natural Limits.
Reflections on Limits of this Kind.*

LIKE every thing else in this world, the military energies of States have their limits.   If we
consider this point attentively, we shall find
that they diminish in a direct proportion to the
length of the line of operation.

This is a consequence of the principle of the
base : for, if it were not necessary to establish a
base previous to advancing ; if modern armies,
like those of the Romans, possessed the sources
of their support within themselves, or could
easily procure them wherever they went, we
might still see conquerors over-running and subduing entire States : a general might leave his
country or his base far behind him, and, like
another Tortenson, be now in Holstein and

now before Vienna.   But at present we have no
such chivalrous campaigns.   As soon as an
army proceeds to a certain distance, it must
have a new base; for, the further it is from its
base, the more acute does the objective angle
become; and I have demonstrated the disadvantages of such a position.   When the objective angle is acute, the flanks, the rear, and
the supplies of the army, are not sufficiently
secured; and it must retreat precipitately, or
perish by famine, or be at last surrounded
and destroyed.

The agency of military energies, like the
other effects of nature, becomes weaker, then,
in an inverse ratio of the square of the distance;
that is to say, in this particular, of the length of
the line of operation.   Why should not this
law, which governs all natural effects, be applicable to war, which now consists in little more
than the impulsion and repulsion of physical
masses?   If, which I do not doubt, it is admissible in the theory of lines of operation, we
may in future easily calculate the utmost extent to which military success may be carried.
Every Power, then, must ultimately be circumscribed within a certain sphere of military activity, beyond which it must take care not to go.

The further an army advances into an enemy's

country, the more does the number of fighting men diminish, on account of the posts it must leave for the security of its rear and flanks. On the contrary, that number with the enemy continues increasing, in the same proportion; because, as he falls back, he approaches his metropolis, and, consequently, the sources of his strength.

It is the same thing, under these circumstances, with the materials of war. They decrease among the assailants, and, on the contrary, become more abundant among those who, as they defend themselves, keep retiring into the interior of their country* For, the former are obliged to collect those materials in the invaded country, or to draw them from their preceding base. Either way, requires much time; whereas, he who defends himself, has, no doubt, united great resources at the

---

* This is true only to a certain degree; and supposing that the army retreating disputes every inch of ground, and has time to carry off from the country it leaves open whatever can be of service to the enemy. It is, no doubt, to be imagined that the defensive, generally speaking, will be carried on in this manner: however, the contrary may happen; and if, by any means whatever, the progress of the army advancing is so rapid and extensive that the other is in a short time dispossessed of a considerable portion of its

central point of his power. He will, therefore, be possessed of formidable masses of
every kind to produce against the enemy, who
has left his own frontiers so far behind him,
that they can be no manner of use to him. It
follows that the State attacked, must inevitably
drive its adversary back to his own base, even
though the latter be the stronger, comparing
the degrees of power of those two States in
every respect.

Nor is it only the length of the lines of operation that weakens offensive operations; the
nature of the territory through which they
pass, contributes considerably to produce that
effect. If lines of operation run over mountains, they must, on account of the time required for the passage of them, be considered
proportionally longer, according to the number
of windings they contain; we know that it
requires more time to go up and down hill,
than to go over an equal space, on level

territory, and without having been able to take the precautions I mention, there is no doubt, unless we suppose the
country conquered entirely ruined and destitute, that the
resources of the army advancing will increase, as those of
the other decrease, or will at least give it the time necessary to establish a new base, and bring its magazines nearer
to it.—COMMENTATOR.

ground; and, what is of still greater consideration, the roads over mountains are usually so
narrow, that it is hardly possible to march
along them more than one at a time. In fine,
mountains are not passable at all seasons; and
if it becomes necessary to go round about, all
communication with your magazines is lost
An army, in this case, may be compared to a
river, of which the source is stopped. The continuance of this situation, even but for a few
months, would be extremely dangerous; for
the enemy, having in his power to reinforce
himself constantly in his own country, would,
in the end, completely ruin an army, which
had so imprudently laid itself open to his
lashes. It must never be forgotten for a moment, that an army, which does not daily receive all kind of supplies, is inevitably going
to wreck.

Mountains interrupt lines of operation in
proportion as they are high, winding, and
steep. A line of operation, carried through such
a country, ought then to be considered, as
being longer than another of equal length,
carried through a plain. In this point of view,
mountains should be placed among natural
limits. I call natural limits, those beyond
which a line of offensive operation cannot

succeed, that is to say, if properly resisted. Water, when it forms a mass of a certain importance, becomes, likewise, a natural limit: this of course is the case with the sea.   Perhaps, however, it may be said that I am wrong in insinuating that the sea obstructs a line of operation, since, on the contrary, both men and provisions are more easily conveyed by water than by land: but the reasons are of different kinds.   In the first place, the sea being a dangerous element, vessels are not only uncertain in their arrival, but are liable to be dispersed by storms; consequently, an army, the reinforcements and provisions for which are expected by sea, can never calculate on receiving them at a given time.   In the second place, let it be observed that, in general, it is impossible to collect in vessels as many men and horses, and as much ammunition, as may be transported at once by land.   Besides, though the sea be the most expeditious way, when the weather is favourable, contrary winds may detain the convoys much longer than they would be travelling to an equal distance by land.   I own, however, that this observation only applies where the country to be invaded is not at an extremely great distance; for, unquestionably, troops may be sent to a grea distance, to America for example, much more speedily by the means of navigation, than they could be, supposing the space between that quarter of the world and us were a continuation of the continent; but, on the other hand, an army would be conveyed from Calais to Dover sooner, were the channel filled up, than can be done in the present state of things.

We see, then, what impediments a simple narrow strait may throw in the way of military operations; and that they are proportionally greater than arise in expeditions beyond the seas.

It is the time consumed in embarking and disembarking, particularly horses, that renders maritime convoys so slow in comparison with those on land, when the distance is moderate Another very great disadvantage to which an army landing immediately on an enemy's territory is exposed, arises from its being compelled to act without delay from the centre to the circumference, that is to say, eccentrically. I have shown the danger of eccentric or divergent operations, a consideration from which it may be fairly concluded, that it is impossible to make a successful landing, unless a party has been secured in the enemy's country to receive

you with open arms.   Thus, the sea interrupts
lines of operation ; and if the continental Power,
determined on sending troops across the sea, have
not the means of carrying at once, for the accom-
plishment of its projects, a much greater share
of the requisites of war, in general, than the
maritime Power against which it acts, it stands
a chance of failing : the sea then constitutes a
very good natural limit *

Lakes, such as those in Canada, which are at
once long, broad, and deep, interrupt, like the
sea, lines of operation, and form excellent natural bounds.   Some have been inclined to deny
this advantage to rivers, but wrongly, in my
opinion.   Great rivers, such as those that receive many other streams in their course, obstruct lines of operation, which in their progress
fall in with them, augment in proportion to
their breadth, depth, and rapidity, the difficulties
of those lines, and consequently form a natural
limit.

It is indeed requisite that a river, to be considered as a natural barrier, should be of sufficient
length : for, if you can turn it, it becomes of no
importance.   With respect to the breadth, it

* I beg to refer the reader to the observations which I
have made in the Preface respecting the probable object
and issue of an invasion of this country by the French, if attempted.—COMMENTATOR.

seems to me that it ought to be of an extent to which musketry cannot carry with effect. There are, however, exceptions: as for example; a torrent of extraordinary depth and rapidity might, on account of these qualities, serve as well as a river, for a natural barrier to a State, in a military point of view.

For, in such a case, neither troops nor supplies of any kind could be conveyed across, except in boats, or over a bridge. The former of these modes is very long, on account of the embarking and disembarking: this excepted, it must be owned, however, that it is much easier to cross a river in that way, in the presence of an enemy's army, than over a bridge. This Charles XII. taught us by his passage across the Dwina, which was a masterpiece of military skill. By means of the boats keeping abreast of one another, it is possible to land on the opposite bank, with a more extended front. These boats, it may be said, might be destroyed by the enemy's cannon; but a bridge, too, might be shattered to pieces by them, and, indeed, I do not comprehend the possibility of establishing a bridge under an enemy's fire: whereas, if you had boats built like rafts, or well lined with bags of wool, there would be scarcely any thing to fear from the fire of the artillery. They should be so formed too, that the troops might

land in a line, and not have to form it. In
this way it would be practicable to pass a river
successfully, in face of the enemy, at least in
the night time.

But even if a river only increases the difficulties of convoys, and renders the connection
with the bank from which you came, and on
which your base is established, uncertain, this
would be a sufficient reason for considering it
as an obstacle to offensive operations; for,
though you should have passed the river in
boats, you must still have a bridge for the
greater facility of communicating with the
other bank. But what inconvenience must attend this communication! A bridge is narrow;
the number of troops and quantity of baggage
that can pass at a time must be small, and
great caution too is required. The rising of
the stream increased by rains, its rapidity, the
masses of ice it carries along at certain seasons,
may all contribute to destroy a bridge, by
which the means of supporting an army is cut
off, and the army placed in the most dangerous
situation.

In fine, should the enemy direct an enterprise against this essential point, the
army which has penetrated into his country
must suspend its progress, must change its

offensive into defensive war, and, if it has to do with an enemy skilled in the art of war, it will be exposed to an attack the more unfavourable to it, as the army so placed acts eccentrically; whereas, on the contrary, the enemy's lines of march and lines of fire will be concentric.

The difficulties of the invading army will be still further increased, if on the opposite bank there are fortresses that must be besieged before it proceeds. As it must receive every kind of ammunition by the bridge it will have thrown across the river, to what inconveniences is it not exposed, particularly when we consider the means the enemy may make use of to destroy that bridge!

And even were there no fortresses to be reduced, offensive operations beyond a river are always very difficult, unless you war with a Power infinitely weaker than yourself. For, if the enemy cannot hinder you from passing the river which separates you from him, he will, at least, find it easy to collect all the requisites of war in greater abundance in his own country, than you will draw from yours. He will check you in front, and also act with advantage on your rear. He will force you to a precipitate retreat towards the only point where you are able to make one. It is easy then to cross a river, but very difficult to maintain a footing in the country beyond it, if you meet with a proper resistance.

If, to obviate these dangers, you separate, so as to prevent your being turned, you will fall into the fault of parallel operations. By dividing your forces so much, you will weaken yourself in all points, and expose yourself to be beaten in detail. But even supposing that you act according to all the rules of art; admitting that, in penetrating into the enemy's country, after crossing the separating river, you carry such numerous corps that the enemy will not dare to come between two; allowing, lastly, that you occupy all the important points, and that you throw such a number of troops on his wings, that you need not apprehend being surrounded, you would, after all, have to fear the superior effect of the greater masses which the enemy may bring against you: for, in the same space of time, he will collect on his side of the river more military materials than you will obtain from the other; first, on account of the length of your line of operation; and secondly, on account of the delay that must necessarily take place in the passage of your reinforcements and convoys over the bridge. The nature of the element to be crossed may likewise.

P

create uncertainty in your communication, independently of the efforts of the enemy. It
follows then that, in respect to time, a river
lengthens a line of operation, and consequently
impedes offensive operations.

Taking all these circumstances into consideration; namely, that it is difficult to pass a river,
and much more so, after passing it, to maintain
a footing in a country where you are opposed
by an enemy who understands the art of war;
that the difficulty becomes almost insurmountable, if you are under the necessity of besieging
fortresses on the opposite bank; that a great
river cannot be turned, and that this impossibility is not to be remedied by attempting to pass
it near its source, because you then meet with
mountains, which present impediments of a different kind; reflecting, I say, on this subject
according to all its data, it must be granted that
rivers, as well as mountains, form natural limits
to States. But it is necessary to observe that
we here speak only of natural, political, and
military limits. The separation caused by a
river has, no doubt, less influence on the nations
that inhabit its different sides, than that caused
by the interposition of mountains; for, the numerous advantages it affords are common to the
people on either bank: but, what is favourable

to social and commercial relations, may impede operations of war, and combinations of politics. Therefore, we may be confident, that the different States of Europe will some day or other extend themselves to those rivers and mountains which, in this point of view, are their natural bounds. That they have not yet done it may be ascribed to the nature of things, which take the order they ought to have, slowly and insensibly; and to the tardy progress of the modern system of war, which has but very lately acquired its present degree of maturity.

Natural barriers require to be further fortified by art; and where there are no natural ones, it becomes necessary to substitute those of art entirely. The frontiers of France, from Landau to Dunkirk, were invincible.

Fortresses may be erected every where, and it is ridiculous to think that they are useful only in places where they are surrounded by natural obstacles which impede the enemy's approaches. On the contrary, the more flat and naked a country, the more difficult is it to approach a fortress which commands it. Still, it is more important to establish fortresses on rivers, and on sites where they defend passes.

Another error, which cannot be too much

exposed, is, that fortresses should not be con-
structed on too large a scale, as they require
immense garrisons. So far from this being
right, a small fortress is of very little use, because you can have only scanty magazines in
it: besides, however strong a fortress of a mid-
dling size may be, the enemy will not be afraid
to leave it in his rear, for he can have no apprehension of the small garrison it contains.
For the contrary reasons, he will not treat a
great fortress with contempt, but be curbed by
it: and if you have the misfortune to be beaten,
such fortresses become a shelter, which enables
you to repair your losses, without fear of being
forced.

CHAP. V,

The Superiority given by the modern System of War to Number over Valour and intrinsic Merit, is favourable to defensive War, and to the Insurrection of a People though opposed by a regular disciplined Army.

THIS proposition has already been nearly demonstrated by what has been said before; for, since masses augment in proportion as they are more compressed towards the centre of the power of a government*, they must necessarily, when properly put in motion, repel an enemy who has penetrated into the heart of a country. Of course, the State that defends itself has always a base at home; while the invader, on the contrary, is ever removing to a greater distance from his; and the more he advances, the more acute is the objective angle within which his operations are carried

* See the note, page 200,

on*. For this reason it becomes easier and easier to manœuvre on his flanks and rear, and to cut off his convoys.

We see, then, that defence is easy against an attacker, whose operations are carried on at a great distance from his natural limits†; and if defensive war has hitherto been considered as very difficult, it was owing to the constant practice of attempting to stop the enemy's progress by parallel movements and well-chosen positions; whereas, on the contrary, a defensive army should never cease making offensive movements and formidable diversions against the enemy: for, a general may, and ought to undertake a thousand kinds of offensive operations in a defensive war, and yet the nature of the war

---

* What the author says is very true; yet I must observe, that as the army which advances will, no doubt, settle new bases in proportion as it proceeds into the enemy's country, and so that each new base shall be supported by the preceding one, the inconveniencies inferred of operations carried on in an acute-angled triangle are less in this case than where lines of operation proceed from a single base.

COMMENTATOR.

† Notwithstanding the observations I have made in the preceding note, and in that in page 200, I am not the less convinced of the many advantages possessed by an army over an attacking enemy, whose operations are carried on at a great distance from his boundaries.—COMMENTATOR.

may not be changed on that account. This kind of war must, no doubt, have appeared the more difficult from the mode of conducting it, as it may even be declared that success in it, according to that mode, is impossible against a skilful and enterprising enemy; for, as I have already observed, there is no position from which an army may not be driven, if the adverse general manœuvres with ability on its rear and flanks.

In examining the proposition; that a people, unorganized but armed, may triumph over disciplined and regular troops, provided the multitude be guided by an intelligent mind, we see one consequence of the superiority which the new mode of war has given to number over skill and tactics. I do not consider my assertion as perilous and rash; on the contrary, it is doing good to point out danger, that the necessity of providing against it may be known.

In a great town, a garrison attacked by the inhabitants armed will be destroyed: the troops have no alternative but to leave the town and bombard it; or, if it be a fortress, to go and cannonade from the ramparts. The soldier, if posted in the streets of a town, will be assailed from the roofs and windows of the houses, and lost; he cannot remain there: nor is he much

better off in the squares surrounded by houses.
The example of Warsaw, that of Ghent and of
Brussels, in 1789, sufficiently demonstrate the
truth of what I advance.

An army scattered in garrisons will have all
its communications cut off, if there is a general
insurrection in the country. If the troops leave
the garrisons in order to assemble, they will be
surrounded on their march, fusiladed if the
country is enclosed, and forced to lay down
their arms. Each garrison by itself is a whole
which, being without a base, can only operate
eccentrically. Incessantly harassed by a multitude of *Tirailleurs*, it must yield to the superiority of numbers *.

* The author is here strangely misled by his theory.
Had he ever taken an active part in a war against insurgents, he would have been fully convinced that a firm government, with an army that remained faithful, will always
quell every kind of popular commotion, however serious it
be. A people in rebellion, left to their own strength, are
nothing more than a mass of men, ill armed, ill organized,
and without discipline, embarrassed even by their numbers,
and who, acting without principle, method, or plan, will
daily commit serious faults, of which soldiers regularly bred
will of course take great advantage. Even supposing the
people to be guided by some able leaders, the operations of
these leaders, however well combined they might be, would
generally fail, because they would not be supported in the
execution of them by men capable of giving effect to their

It may be said, that this might happen in quarters where the troops are cantoned, but

designs. Besides, the insurgents would have no artillery or cavalry, and frequently no ammunition; they can have no magazines, hospitals, nor any kind of fixed establishment: the consequence of which is, not only that they will not have it in their power to pursue any temporary success they may chance to gain, but that the slightest defeat, such as would be nothing if experienced by the army, will occasion a complete rout to them, and that the more dangerous, from their having no safe points of support in their possession.

If it be said that they would live on the provisions they would find in the country; I answer, in the first place, that they could not do it without separating; for, if they concentred themselves, in order to oppose a sufficient force to the troops sent against them, there being a great number of men assembled at places of little extent, it would be impossible that those places should supply their wants; and, if they separated, becoming every where feeble, they would infallibly be destroyed party after party. In the second place, if the insurrection lasted any time, agriculture would be necessarily neglected; for the people could not persist in fighting, at least in a manner from which any decisive result might be hoped, and at the same time cultivate the fields. It must follow, then, that, after a certain time, the insurgents would be starved, and forced on that account to submit, even if they were not reduced by force of arms.

But if there were a famine in the country, what would the government do to subsist its army? I reply, it would subsist it from the magazines which it would, no doubt, have

1

never in fortresses where they have it in their
power to fire from the ramparts on the town.

established on account of the insurrection; and this with the
greater ease from having in its hands at the time, and un-
shackled, all the pecuniary and other means of the State.
It would afterwards subsist it, if driven to this resource, by
means of foreigners, who would supply it even on credit
with all it might want. It would be a vain objection to say,
that the insurgents would obstruct the communications, and
either prevent the entrance of the provisions or seize them;
for, besides that it is physically impossible that an insurec-
tion, unsupported by the army, should extend to all the
points of a State, the troops would find means to open the
passages necessary for the arrival of the convoys, and to secure them. Some countries, I grant, such as Swisserland
and the Low Countries, succeeded, in times past, in rendering themselves independent; but, besides that armies were
then much less numerous than at present, and consequently
garrisons much weaker relatively to the population, it is
likewise to be observed, that the great distance at which
those countries were situated from the centre of their government was much against the conveyance of the necessary reinforcements in time. Let us cast an eye on all the insur-
rections which have since taken place, and which have not
been agreed to by at least the greater part of the army, or
supported by foreign troops: is there a single one that has
terminated in favour of the insurgents, or that has not even
been quelled in a very short time, and almost without striking a blow? There is, no doubt, that a State may lose its
distant possessions by a popular insurrection; and for this
reason, that it either has not troops enough on the spot, or is

But it is usually in those important places, and in capitals, that the magazines, and all the provisions of war, are collected: now, capitals, like large fortified towns, contain a considerable population. In a popular insurrection, then, the number concerned in it may be so great as to compel the soldiery to keep together in their barracks, where they may be blockaded, starved, and at length constrained to surrender with the magazines which constitute the strength of an army *.

prevented by distance from sending sufficient reinforcements thither in time : but a government in the centre of its power has certainly no such danger to run.

I say nothing of the divisions which are the natural and indispensable consequences of the ambition of the leaders; of the mutual jealousy they excite, nor of a thousand other circumstances, which all necessarily tend to the certain annihilation of the insurgent party. Besides, against this party there will be another in the country itself, independently of the army; for, even were all the inhabitants to think alike, which it would be absurd to believe, their interests would be different.

It is evident, then, that an insurrection, not supported by the army or by an adequate number of foreign troops, can produce no other effect than that of a devastation of the country where it takes place, and a massacre of some of its inhabitants, useless to the insurgent party, which in the end must undoubtedly be crushed.

* We must suppose a government very passive or ill-

Among the ancients, with whom superiority
of discipline, bravery, and a knowledge of Tac-

informed, to imagine that it had not at least some notice of
a considerable insurrection while plotting; an insurrection
of this magnitude could not break out unexpectedly: it
must necessarily have some prognostics, and whether the
government holds the clue, or only has some notice of
it, surely it could not have neglected to collect about the
metropolis and fortresses containing its magazines, a sufficient number of troops to protect such important points:
now, if these precautions are taken, I not only think
that the troops cannot be forced to surrender, and to give
up the magazines they were stationed to guard, but I am
persuaded that most frequently they would not be attacked,
and that the insurrection, if it does break out, would soon
be reduced to insults and insignificant clamours.

The author adduced the examples of Warsaw, Ghent, and
Brussels, to prove that the garrison of a great town must be
overcome if attacked by the inhabitants armed. But I
would ask him, in the first place, whether he be very sure
that the garrisons of those towns were not much too weak,
compared to their size and population? In the second place,
whether their evacuation were not rather a measure of prudence than the result of necessity? and whether that measure
were not determined by reasons absolutely unconnected
with the resistance which the troops might have made? I
would ask him, likewise, if those troops were surrounded in
the course of their march, fusiladed, and compelled to lay
down their arms? I would ask him, lastly, if, after quietly
assembling and waiting some time for the reinforcements
they wanted, they did not return and put an end to the in
surrection throughout the country, and take up their quar

tics, always ensured the victory, a disorderly multitude could do nothing against the smaller number who acted in a regular manner. The effect now produced by fire-arms was unknown: the weak could not with impunity defy the strong, nor the coward insult the brave. At present, the number of shooters must sooner or later decide the victory, provided they be organized and distributed so as to fire with ease. We have seen too that, among the moderns, an army attacked on its flanks, or whose magazines are threatened, is in imminent danger; circumstances favourable to a people in insurrection, against a regular army: a people in this state wants only arms and intelligent leaders. It is the consciousness of the strength of a nation, in such a case, which apparently deters many governments from arming their subjects; but certain States have been compelled to it, in order to increase the number of soldiers absolutely want ed, and hence the origin of levies in mass.

ters peaceably in those very towns, the temporary possession of which gave the insurgents no other advantage than that of prolonging, for some months, the troubles and the misfortunes of their country ?—COMMENTATOR.

CHAP. VI.

As, sooner or later, the various States of Europe will extend themselves to their natural Limits; and as it is both Fruitless and Dangerous for a Government to carry its Operations beyond the Frontiers prescribed to it by Nature ; the necessary Consequence of this Order of Things will be perpetual Peace*.

WE must suppose men struck with an inconceivable blindness, if we can persuade ourselves that they will still go to war, when this state of things takes place; for, the object of the continual wars we have at present, is aggrandizement by conquests: but, if experience demonstrates the impossibility of attaining this object, will not men give over fighting? There are certain Powers that maintain a constant state of war, because they have not yet attained the extension they ought to have, and because it is necessary that they should ag-

---

* I have stated, in the Preface, my opinion relative to what the author says in this Chapter.—COMMENTATOR.

grandize themselves to be able to resist others; well knowing, as they do, that they are exposed to be crushed by superior masses, when those masses are not stopped by obstacles interposed by nature.

Thus, the sooner Europe is divided into Powers enclosed on every side by natural limits, the sooner will a state of perpetual peace be established in this quarter of the world. It is evident, then, that the wishes, and the exertions of the friends of humanity, should tend to consummate, as speedily as possible, this salutary event.

When that time comes, the heads of States, convinced by experience of the inutility of their attempts at aggrandizement beyond the bounds fixed by nature, will moderate their ambition, and will have no longer any pretence for keeping up those great armies so injurious to society, and which will then be confined to what ought to be their only use, namely, the maintenance of the police, and of order in the interior of a State. The governments which, in spite of this change in political order, shall keep on foot the exorbitant number of their hirelings, will prove that they are the enemies of their respective nations, and that they dread an insurrection against the tyrant.

Uninterrupted peace will considerably in crease the physical welfare of men; for, war is the most voracious consumer of the elements of existence. The number of producing agents being diminished, the quantity of produce must necessarily be so. Never was there a more unfounded objection than, that from a continual peace we should have to dread an excessive population. The more men there are, the more productions; and in a depopulated country, one runs a risk of starving from a want of consumers.

At all events, there is very little fear of this pretended danger in Europe, so poor in men, and so thinly inhabited. How many deserts does it contain awaiting the hand of the cultivator! This is not the place to speak of the causes, which, by creating a thousand difficulties in the means of living, generally render in these countries the lot of the multitude so wretched. Let me be permitted but two observations: there is more ·space in Europe, supposing it equally divided, than can be cultivated by each individual, or than is necessary for his maintenance; and every body knows that a garden yields more than a field of the same size, because, the produce arising from intense cultivation increases in a ratio in finitely exceeding that of its extension; just as a profound thinker, who considers a single object, unfolds and illustrates many more ideas than the superficial man, who thinks of several things at once.

Morality gains by the ceasing of war, that pest of the virtues. I shall not only say, how much does the habit of murder and robbery debase men to the level of brute beasts; how much does the daily sight of the horrors of war stifle sensibility in the best hearts; but, let us attend to the abominable opportunities afforded by war, of men's enriching themselves by plundering, not only the public, but the wounded and the sick who are left to perish in protracted torment, for want of the necessary means of cure; and we shall be convinced that wealth, which should only be the recompence of the useful citizen, frequently becomes, by war, the portion of men the least worthy of it.

The blessing of perpetual peace would put an end to the error, that the science of war, that science of robbery, rather than of murder, leads more than every other to immortal glory. Not only the extinction of war, but even a more general knowledge of its principles, would contribute to dispel this delusion; and, when

Q

experience has shown that ability can do no thing against the shock of superior physical masses, the estimation of this art, and of those who profess it, will fall rapidly.

I call war the science of robbery, and not that of murder, as it has been hitherto denominated; because, to rob is its object, and to kill is only a means. It seems that killing was thought something more noble than forcibly taking away what belongs to another, when men applied to the art of war the denomination I mean now to deprive it of. There have been, I allow, some ardent warriors, who have made war for its own sake, and for the pleasure of fighting and destroying, such as Alexander and Charles XII.; but these examples are very rare, and cannot overturn my opinion as to the motive by which warriors are usually actuated.

The corruption of men engendered war, and war in turn became a stimulant to the corruption of humanity. The extinction of the one would be followed by that of the other, at least in great part. Such will be the blessed con sequence of the perpetual peace, for which we shall be indebted to the mode n system of war, the result of the invention of gunpowder; of that invention, which has been cursed a thousand times as a scourge of mankind. It is to this

1

discovery we owe the principle of the military base displayed in this work; and this principle, approved and confirmed by experience, leads to perpetual peace.

I believe I am the first person who has started this idea. Where could there be any natural bounds for a Roman army marching on without a base and lines of operation? What had those excellent troops to fear from numbers at a time when the power of armies depended upon their inherent qualities? Were there any obstacles to their endless triumphs, and consequently to the perpetuity of war? Then the world might be conquered, whereas the balance of Powers is the happy result of the system adopted by the moderns. Considered as an art, war was infinitely superior among the ancients to what it is among us; because physical masses were of no value. But the new system excels the ancient in effects beneficial to humanity, and the superiority of the latter consisted only in the dreadful evils it produced.

The nearer the new system advances to perfection, the more it differs from the ancient. Since the invention of gunpowder, all history demonstrates this truth to us; and from that period the art of war, considered as an art, has been constantly sinking; because, ever inadequate to accidental events, it has become more and more dependent upon them.

In the operations of war, as in every thing else, the overcoming of difficulties has always been considered as a mark of genius in him who succeeded. What will be the case, when insurmountable difficulties multiply, and that in proportion too to the improvement of the art? Will not the consequence be that the sphere of military genius will at last be so narrowed, that a man of talents will no longer be willing to devote himself to this ungrateful trade, but prefer employing his powers on objects of more general utility!

War will be no longer called an art, but a science; for, art is the application of science. Science is in the mind only; art descends from the mind into the sphere of activity. Art is all that can be done, whether good or bad: these qualities accord not with science; we know it, or we do not know it. We say of a science, that it is true or false; of an art, that it is good or bad.

Thus, the more the sphere of the *art* of war shall be contracted by the increase of insurmountable difficulties, the more, on the contrary, will the *science* of war be extended; and when it has attained its highest degree, it will finally reduce to principles the possible and impossible of this kind. Every one will be then capable of understanding the application; the art itself will be a science, or be lost in it. But it will be difficult for a man to distinguish himself in what will be so common; consequently the pas sion for military glory will be extinguished, and thus perpetual peace be the better established.

CHAP. VII.

THE CONCLUSION.

TAKING a review of our inquiries in this division, we shall conclude: " That the num-
" ber, and not the excellence of troops gives
" success in the modern system of war; that
" in future, small States will not be able to
" conquer great ones; and that, for this rea-
" son, Europe will in the end be divided
" among great Powers, which, however, will
" not pass their natural bounds; because be-
" yond those bounds, offensive operations will
" no longer be effectual; whereas at home de-
" fensive war will be easy and prosperous.
" We shall conclude, that a people armed
" will henceforth be able to overthrow a re-
" gular army; and that perpetual peace will
" be the fruit of all these circumstances toge-
" ther. Lastly, the most positive conclusion
" is, that humanity will find its future safety

" in the necessity of a base to lines of mili-
" tary operation, and that this necessity will
" daily impress itself more and more upon the
" mind."

END OF THE SECOND PART.

Q 4

THE

SPIRIT

OF THE

MODERN SYSTEM OF WAR.

PART THE THIRD.

APPLICATION OF THE PRINCIPLE OF THE BASE
TO PAST MILITARY EVENTS, AND TO THOSE
THAT MAY TAKE PLACE IN FUTURE.

CHAP. I.

*Of the Period at which the first Developement of
the Principle of the Base was made.*

I HAVE said more than once, that the military operations of the ancients required no base. A Roman army, containing within itself all its means of support, was a body of the highest degree independent of every external influence: it was a moving magazine. Full of confidence in its moral and physical strength, and never doubt-

ing of victory, it was little afraid of being surrounded.

We cannot say so much of the Greeks: however, the smallness of their armies, the petty
number of their cavalry, and their temperance,
contributed to render magazines little necessary
for them. The oriental nations had scarcely
any mounted troops, some few excepted, as the
Parthians, who always fought on horseback;
but they were light troops, that had no occasion
for storing forage in plains always green: as
we see at this day the Tartars subsist without
such precaution, because their warlike excur-
sions are of short duration.

The nations that overturned the Roman power
appeared, as the Tartars, in troops of light
horse; or almost without cavalry, as the Germans and Franks. From this very circumstance
they stood in no need either of magazines, or of
lines of operation, or of base.

In the middle ages, which followed the fall
of the masters of the world, war degenerated
into freebooting carried on by parties of troops
on horseback: valour then consisted in being
cased in iron and keeping a firm seat in defiance
of the adversary's shock We find indeed a
numerous infantry re-appearing in the croisades,
but no system for supplies: Mahomet II. at the

siege of Constantinople, was the first who made use of heavy pieces of artillery The example was followed by others, but cautiously, and much time elapsed before the system of fire was brought to any perfection. The Turks seem to have made the first progress in this system ; for, under Soliman II, their infantry was the best in Europe, though they afterwards lost their superiority. There was still, during the *thirty years* war, a wavering between the old and new system, nor did the genius of Gustavus Adolphus decide for either. We find no trace of a regular system in that memorable war. The armies were small and lived by plunder, and of course never was there a war more marked by devastation. Gustavus Adolphus flies from Pomerania into Bavaria; he hastens from thence into Saxony; he encamps with a river in his rear. Tortenson overruns Germany from one end to the other. The Swedes are now on the Rhine, now in Bohemia, now in Lower Saxony. Veimar made war like a knight-errant; there was no system, no order, no object; it was every where a chaos.

Amidst this confusion Mareschal Turenne was the first to introduce regularity. We discover the system of the base in the admirable campaigns of 1674 and 75. These are the first

campaigns of modern history worthy of notice.
Previous to these, generals obtained celebrity
by some brilliant feats, entirely unconnected,
however, with a chain of action. Montecuculli
and Turenne, both masters of the art of war,
were the first who gave the world an example
of a systematic campaign, designed and conducted without an error.

The system of fire was brought to great perfection in the reign of Louis XIV. Vauban
created the Tactics of sieges. The most important change that took place, and of which all
the rest were but consequences, was the aboli-
tion of pikes and the substituting of bayonets
for them : from this the old system of steel arms
received its last blow, and the foundation of all
the improvements since made in the fire of Tactics was laid. That epocha should be considered as one of the most important in the history
of the art of war. It is from it that the modern
system is properly dated; for, a change of arms
must necessarily induce a revolution in the
military art.

In the *succession* war we observe marks of
the necessity of a base beginning to be felt.
Prince Eugene and Marlborough took fortresses
on their flanks, before they advanced further
Thus, when they sat down before Landrecy

they were in possession of Lille, St. Omer, &c. Their base however was not sufficient, otherwise the battle of Denain would not have overturned all their projects. Charles XII. was absolutely ignorant of what a base was. His military career may be compared to a literary work containing beautiful unconnected parts, but which forms no whole regulated upon a plan. His military feats are proofs of an astonishing warlike genius, but all his wars are the chivalrous campaigns of a novice. Every military man must pay a tribute of admiration to his passage of the Dwina, his battle of Narva, and his battles in Poland. His very adventure at Bender deserves to be studied by commanders of parties. How many resources did he not display in fortifying and defending his house? But his marches, uncertain and without a plan, and especially that into the Ukraine by the advice of a Cossac, are beneath criticism. He lost his army, his glory, and his fortune, in the plains of Pultawa, because he led his troops thither without a base.

The war of the succession formed a man of genius in the military profession, who contributed more than any general to the perfection of the fire of Tactics; for, without him, Frederic II. himself would have been nothing in that respect.

Thus, in examining the consequences of the modern system of war, the Prince of Dessau was the principal instrument of the most important revolution presented in the annals of the world; and the two inventions of this officer, which have had such powerful effects, are the *iron ramrod* and the *equal step*.

The iron ramrod gives an activity to the fire of musketry, which has rendered the effect of it much more terrible; for Folard, who was unacquainted with it, speaks with contempt of the musket fire, such as it was in his time. The equal step must likewise be considered as one of the causes by which fire has been improved, because, to produce an effectual one, it was necessary to know how to make movements in lines of great length. The Romans seem to have been acquainted with the equal step, which was forgotten in later times, like almost all the inventions of antiquity. The Prince of Dessau did not recover it, for he was too ignorant to doubt the military art of the Romans; but he invented it. It was he, too, who first thought of drawing up the infantry in three ranks only, instead of four. All this contributed to the perfecting of fire; and Frederic II. assuming the ideas of the Prince of Dessau, improved upon them. Since that monarch, the Tactics of the infantry may perhaps have received a modification more analogous to the new system of war, but the foundations of the present military art are due to the Prince of Dessau.

The infantry formed according to the precepts of the Prince of Dessau was rendered lighter by Frederic II.; but at the same time the fire became more irregular. He diminished the internal strength of the infantry, but taught it to outflank the enemy's wings with greater dexterity. His infantry was no longer seen keeping up a platoon fire with such regularity as at the battle of Mollvitz, but his columns deployed with greater rapidity, and in general he introduced among his troops that promptitude, which was the cause of all his successes, and which gave the new system a further advance towards its maturity

The Tactics of close infantry must necessarily have attained this point, previous to the last modification of the system of fire, that is, to the art of *Tirailleurs*, which, as I observed, is more agreeable to the military genius of our times. All the productions of the human mind are successive. Yet, let what will happen, it is impossible that the close, or as it is also called, heavy infantry can ever be dispensed with. We do not find that Frederic conceived, or even followed the principle of the base in the two first wars in Silesia. That monarch did not commence his military career so brilliantly as the great Condé, though he afterwards excelled him. The battle of Mollvitz was gained immediately after the king had left the army, and the Prussians owed that victory to the iron ramrod, which doubled the activity of their fire. The equal step contributed likewise to their success; for, their infantry were as orderly as on a parade, by which the Austrians were disconcerted. We find a disposition still more orderly at the battle of Czaslau than at that of Mollvitz, and particularly a more active cavalry. Frederic was the creator of his cavalry, and not of his infantry, as has been erroneously imagined. He brought the cavalry to greater perfection than any other general. Before his time the squadrons never went out of a trot, and used firearms more than their sabres. Frederic was admirably seconded by great cavalry officers, such as Generals Golz and Seidlitz, and no cavalry in Europe has yet equalled the Prussian cavalry, for great movements.

However, Europe was not dazzled by any distinguished military talents in the first war of

Silesia. Frederic, in his history, blames Mareschal Schwerin for persisting in covering Upper Silesia, and for requiring succours only for that purpose. To this the king attributes all the retrograde movements which preceded the battle of Mollvitz: but it is probable that, at that period, Schwerin was a better general than his master, and that he was perfectly right in his conduct.

In the second war of Silesia, Frederic, as he confesses himself, could not keep Bohemia, which he had conquered very rapidly, against the superior ability of Mareschal Daun: but, besides this, the nature of the country is one cause why the Prussians will never be able to maintain themselves easily in Bohemia during winter.

The chain of winter quarters there are not protected by any river, and the mountains which separate Bohemia from Austria are not considerable enough to constitute a barrier. Besides, the chain which the Prussians would endeavour to form in Silesia is too short, and has Moravia behind it. From this it appears that, in order to cover Bohemia, the Prussians would find it necessary to make themselves masters of the fortresses in Moravia, or rather to penetrate as far as the Danube.

R

To this is to be added, that there are loftier mountains and more difficult defiles in that part of Bohemia which faces the Prussian States towards Saxony and Silesia, than in the part lying towards Austria and Moravia. This interrupts the communication, and renders conveyance more difficult. Lastly, it must be remembered, that in Bohemia, the Prussians are very far from the centre of their power, and the Austrians very near theirs; whence it follows, that the latter can collect a large quantity of the materials of war in a shorter time than the former, and must, consequently, force them to fall back

In 1745, the Austrians penetrated into Silesia, whither Frederic II. confident of the superiority of his troops, had skilfully allured them. It was at the battle of Friedberg that he first displayed the system of the oblique line. But, independently of excellent dispositicns, the bravery of the troops contributed greatly to the gaining of that battle. One regiment of dragoons took seventy standards from the enemy.

At the battle of Sorr, Frederic was surprised in his camp. His army was saved from total ruin by the extraordinary promptitude with which he formed his cavalry, and by the presence of mind and ability of General Golz, who, by means of that cavalry, threw the enemy into confusion, and put a stop to his success. Prince Charles had conceived the plan of this attack like a master of the art, but the execution of it was pitiable. Be that as it may, this battle shows to what a degree of perfection Frederic had already brought his cavalry, in the second war of Silesia.

The battle of Kesselsdorf was gained by the Prince of Dessau, who, by a feigned retreat of his infantry, enticed the Saxons from their entrenchments, and his cavalry cut them to pieces. The principle of the base was not yet developed in this second war of Silesia.

I consider the *seven years war* as the true æra at which this principle was adopted and practised, from a full conviction of its importance. Frederic is the author of this revolution in the military art, for he brought the system of fire almost to perfection; though it is true that his improvements are founded on the Prince of Dessau's discoveries. The perfecting the fire of Tactics, evinced the necessity of increasing armies for the purpose of extension; it also caused the augmentation of the artillery and cavalry, for the weakness of infantry was ac-

knowledged.   All this multiplied the wants of
armies; immense magazines became requisite;
and from these various causes united, the principle of the base followed as a necessary effect

CHAP. II.

*Examination of some Campaigns of the seven Years
War; and Remarks on that War, in general.*

BEFORE Frederic could invade Bohemia with
success, it was necessary that he should be
master of Saxony, and that he should destroy
the Saxons before they could be succoured by
the Austrians.   He executed this plan with
such skill and rapidity, that the campaigns of
Cæsar himself, present us with nothing more
brilliant.   The direction of the three columns
with which he invested Saxony, is a masterpiece
of military art; for it rendered it impossible
for the Saxons to take any position in their
country, without exposing themselves to be cut
off from Bohemia, and made prisoners.   Now,
as they preferred making a stand at home, to
flying hastily into Bohemia, they were effectually surrounded and compelled to surrender.

General Tempelhoff, the first military
writer who unfolded the theory of lines of
operation, justifies Frederic against the cen-

sures of some distinguished military characters,
and among others Lloyd, who blames him for
not profiting immediately of his advantages,
after the invasion of Saxony; and for not marching rapidly to the Danube. General Lloyd
is regardless of the calculations which pro-
hibit the removal of an army from its magazines in a given time beyond a certain distance,
under the pain of being exposed to want provisions. Yielding at times to his imagination,
he forgets that those campaigns of giants, in
the present mode of waging war, are practicable only on paper Thus, Lloyd's censure
touches not Frederic It was prudent in him
to continue at that time in Saxony, and to take
advantage of the winter to reinforce his army,
from the many resources which the country
might offer. Every part of this campaign is a
masterpiece; and if Frederic, in his first wars,
appears an unpromising scholar, at the commencement of his third, surpassing himself, he
stands forth a finished master in the art of
war.

In consequence of the situation of Bohemia,
in respect to Prussia and Austria, all the offensive operations of the Austrians against the
Prussians, from that country, must be eccentric, and of course bad. The more defective

will be their base, the more they approach the angle of the top of the triangle, formed by Bohemia against the Prussian States; for, in this case, their base will be less than 90 degrees, and they will, consequently, be exposed to all the disadvantages of operations undertaken in an acute-angled triangle.

But, as the salient angle, formed by the frontiers of Bohemia, lies against Lusatia, I must conclude, notwithstanding General Lloyd's opinion, that it is not by that quarter that the Austrians would be able to direct their operations with the greatest advantage against the House of Brandenburg That point, indeed, threatens the heart of the Prussian monarchy; but the imperials would not reach it. To penetrate to Berlin, they must be masters of the Elbe to Dessau, and, consequently, of Dresden; they must conquer Glogau; and then they could not maintain themselves at Berlin, without being in possession of Magdeburg. In general, when examining military plans, it is not enough to consider action, or pression; we must likewise have some regard to the contrary of them. No doubt, an army may go from Lusatia to Berlin, if it is not opposed; but, in case of resistance, it would be exposed

to be cut off, and be compelled to retreat precipitately.

To have lines of operation proceeding from
a better base, the Austrians may attack Prussia by the middle of Silesia, and of Saxony;
but the mountains of Silesia, and of the county
of Glatz, and the impregnable fortresses in
those countries, are obstacles almost insurmountable: and should the Austrians proceed
at the same time by Saxony, their operations
would be divergent, the bad effect of which
we have already stated.

But, south of the fiftieth degree of latitude,
which cuts Bohemia into two equal parts, all
operations are in favour of the Austrians, not
only because they are there nearer the centre
of their power, but because, from the frontiers
of Moravia, Austria, and the Upper Palatine,
they would surround the Prussians, and put
them into imminent danger of being cut off
from their country and magazines. It ought
always to be with a view to the great principle
of the base, that the interests of States, and the
fate of Nations, independently of all other advantages, should be calculated, and it will be
so, more and more, in future; and, according
to that principle, it may be foreseen that, sup-

posing war between Prussia and Austria, to be
confined to Bohemia, Austria would find it
difficult to keep from her rival, that part of
the country which lies to the north of latitude
fifty; and that, on the other hand, Prussia
would find her enterprises to the south of that
latitude, attended with insurmountable obstacles.

It is only by diversions from Moravia into
Silesia, on Neisse and Kosel, that the Austrians
can defend Bohemia against the Prussians;
consequently, far from censuring the emperor's
generals for leaving an army of twenty thousand men in Moravia, during the campaign of
1757, we must blame them for not leaving a
larger one, as those troops might have served
to facilitate a diversion into Silesia, while the
King of Prussia was penetrating into Bohemia.
The art of diversions is the great art of modern
war, and the only means of success.

Had the Austrians adopted this principle,
they would have forced the King of Prussia, on
the occasion of which we have been speaking,
to put himself on the defensive. Far from
doing this, they committed two capital faults,
which they have since often repeated: the first
was, the leaving of their army in Bohemia,
collected into a mass, instead of spreading their

troops; the second was, the defending of themselves passively, without making an attack any
where. Had their defence been still more vigorous, it would have been to no purpose
against the invasion in four columns, planned
in a masterly manner by Frederic. That invasion was conformable to the principle of the
base. The four columns penetrated at once,
which was infinitely better than advancing
with only one principal column, leaving corps of
defence on the flanks and rear. These corps
lose much of their efficacy, and of their energy,
merely because they are on the defensive.
They are in danger of being beaten, if attacked vigorously, and their defeat is attended
with incalculable disaster to the acting column.

The Austrians suffered themselves to be driven back from all sides to Prague, like a flock
of sheep, instead of making, according to good
principles, an eccentric retreat, leaving Prague
to itself, retiring to the frontiers of Austria,
and, after receiving reinforcements there, returning on their adversaries with new vigour.

They ought not to have waited a battle in
the vicinity of Prague; for they should have
known that, on that day, the Prussians would
be superior to them in force. For that very

reason it was Frederic's interest to bring them
to action.   The battle of Prague was as decisive as any mentioned in modern history.   The
two armies engaged the whole length of the
line.   Among the ancients, the consequence
would have been annihilation to the vanquish-
ed, and Prague would have become the prize
of the victors; but, by an effect of the striking difference between the modern and ancient
system, the Austrians were able, in a month, to
fight new battles.

It is the immense wants of modern armies,
and the danger they run in going to a distance
from the sources by which they may be supplied, that render the bloodiest battles so indecisive in our days.   Tempelhoff has shown,
that an army should not leave, between it and
its magazines, a greater distance than will
allow a new delivery of fresh flour to the
bakers in three days, and the bread-waggons
must be able to go and to return in six.   But
this is the difficulty attending only on the supply of bread; that of forage is still more material.   An army of one hundred thousand men
has forty-eight thousand horses; to supply food
for this number   requires four thousand five
hundred wains.   Add to these two thousand

more for the carriage of flour, which make
together six thousand five hundred; and supposing each of these drawn by four horses, they
make an augmentation of twenty-six thousand
to the number of these animals in an army.
Tempelhoff does not even take these twenty-
six thousand horses into the account; nor has
he observed that, when forage is taken green, the
means of establishing magazines in autumn, are
destroyed.   The result of these calculations
is, that prudence forbids an army to go further
than three days march from its magazines.

The King of Prussia's march against the
French, and the battle of Rosbach, which de-
livered his right flank; his rapid return into
Silesia, and the battle of Lissa, by which he
rescued his left flank, are such masterpieces of
activity and science in war, as will render, for
ever, his campaign in 1757, a model to be co-
pied by all who would extricate themselves
brilliantly from an embarrassing situation.
It is only by a sudden attack that you can
drive a superior enemy from your flanks.   It is
true, that if he is skilful, he may frustrate this
attempt of despair; but that was not the case
in the instance adduced.   The art of war was
entirely on the side of the weak; and the

strong, whose advantage was rendered of no avail by their ignorance, hazarded battles, and were beaten.

We hear of nothing more ridiculous than the conduct of the Austrians, at Lissa. They had a river in front, but crossed it, and left it in the rear, as if on purpose to impede their retreat. They suffered the king to manœuvre at his ease, in front of them, without making a movement. If their whole line had advanced, they must have gained the battle, for they had it in their power to out-flank and surround the two wings of the Prussian army: but they pleased to give the king full time to plan their defeat; an unaccountable absurdity, that will appear incredible to posterity.

The king's dispositions in this battle were admirable, but against so awkward an enemy, the skill displayed was a waste of genius. I do not know, however, whether the battles of Crevelt and Friedberg are not still superior. At Lissa, the king's left flank was not covered.

This campaign, more abounding in battles than any other of the same war, which may be called *the war of battles*, ended with this action, the most remarkable in modern military history. Nor do I know that we can find, in all history, a year, in which so many great and decisive battles were fought.

It remains a moot point, whether, in the campaign of 1758, the King of Prussia, after retaking Schweidnitz, should not have fallen on the Russians instead of besieging Olmutz.

Had he determined in favour of the former design, he should have passed the Vistula, at Warsaw, and attacked the Russians in their quarters dispersed on the Lower Vistula, and in Eastern Prussia. From his known celerity, it was possible for him to have made this attack before the Russians could have assembled. The question at present is, whether he could have easily taken so long a march through Poland, without finding magazines in his way. This difficulty is great; for though thirty thousand men might have been sufficient for that expedition, such a number of soldiers could not be moved in our days, without a considerable train; and though Poland abounds in corn, it is thinly peopled, and provisions lie wide, over a great space of country. I think, on this account, that it would have been better for him to content himself with occupying the western bank of the Vistula, in order to defend the passage of it against the Russians;

and though Poland was at that time neutral, the King of Prussia was not bound to respect a neutrality, which the Russians had infringed. Besides, it is better to make an enemy of a neutral State, when it is without defence, than to allow its neutrality.

If the king had taken possession of Cracow and Warsaw, and fortified those places; and if he had manœuvred so as to make himself master of Dantzick, the Russians, who, from a want of magazines, could not have moved till the end of June, would not have been in a situation to give the slightest resistance to those operations, rapidly executed. Thus master of the part of Poland and of Prussia, situated to the west of the Vistula, he would have had plenty of recruits and provisions at his command. He would have made the Vistula a natural barrier against the Russians, who would have found it difficult to pass it. Brandenburg and Pomerania would have been protected from all insult. Posen and Colberg would have formed a second line of fortresses not to be reduced. Frederic, by thus circumscribing the territory he had to defend, would have shortened his lines of operation, and increased his defensive strength. In all probability, he would have kept the country during the whole war.

He would have avoided the dangerous position in which he found himself, when the Russians and Austrians united their efforts against him, in 1761. He would have spared the dreadful effusion of blood, caused by the battles of Zorndorf, Zullichau, and Francfort. It may even be presumed, that that country would not have been taken from him at the peace, if he had offered Eastern Prussia to the Poles in exchange. The Prussians would, from that period, have become a Power possessing a sufficient base. Such would have been the immense advantages of that operation in Poland, had Frederic preferred it to that march to Olmutz, which owed its success entirely to the little skill of the Austrians in the art of war.

I will not dwell on the other movements of this campaign; but, generally speaking, I will lay it down as a principle, that when an army is obliged to fight a battle, it has certainly committed some previous fault against the present rules of war. The battle of Zorndorf, for instance, might very well have been spared.

The operations of Prince Ferdinand of Brunswick are the most striking features of this campaign. The plan and the execution deserve alike to be admired. What activity, what exactness, what precision in his movements, when in two months he drove the French army, superior in number, from Stade beyond the Rhine! This example proves how dangerous it is to go too far from one's frontiers. What tributes of applause does not the passage of the Rhine, by the same general, merit! What superiority in the skill of dispositions did he not display at the battle of Crevelt! To check, with two small corps, the front of an enemy much stronger, while a third attacked him in flank, is an operation that may be regarded as being planned to perfection. In this war, there is but one battle, which, for the beauty of its plan, deserves to be compared to that of Crevelt, and that is the battle of Friedberg. It was the more requisite for Prince Ferdinand to bring on an action at Crevelt, as he could do nothing against the supplies of the French, who had a sufficient base. However, this battle, like many others, proves how indecisive these general engagements are, in the modern system of war; for, in spite of his victory, Prince Ferdinand was under the necessity of re-crossing the Rhine before the end of the campaign. This Prince, who had under him only raw troops, levied in different States, has shown in his campaigns to what advantage an able general may turn the weakest means.

However, the King of Prussia's famous battles in the *seven years war*, notwithstanding the
fine dispositions which rendered them models,
deserve but to a certain point the admiration
of men skilled in the art of war: for, if they
could have been avoided, as may reasonably be
supposed, they should no longer be considered
but as desperate attempts to extricate himself,
by victory or death, from a perilous situation.

I am convinced that Frederic's campaigns
would have been more brilliant, and more useful to him, if he had always been guided by
the principle, that in a defensive war a general
should constantly harass the flanks and rear of
the enemy, and not engage him in front; and
that, by so doing, he shackles his adversary's
advantages, and renders his superiority of number unavailing.

The Russians committed a great many faults
in this war against Prussia, the worst of which
was that of not spreading themselves on the
eastern bank of the Vistula, and establishing
magazines there; for, by neglecting this precaution, and having no base but on the Lower
Vistula, they exposed themselves to the danger
of being cut off and overthrown as soon as ever
they left that river. They, afterwards, with
the Austrians, were faulty in not sufficiently

spreading their troops, the whole of the war, and in not dividing into several columns, which would have enabled them to draw greater advantage from their superiority.

Prince Ferdinand set the example, before the battle of Minden, of a most skilful distribution of detached corps. The body of the army was divided into two: a large detachment was placed near Hervorden, on the flanks and rear of the enemy, whose supplies were by that means threatened. There was another corps at Lubeck, for the purpose of keeping up the communication with that of Hervorden. All these positions surrounded the enemy. All the lines of fire were concentric. How much must these various circumstances have contributed to the success of the battle!

Mareschal de Contades might have retreated by an eccentric march, that would have proved triumphantly the excellence of this kind of retreat. Let us suppose that he had divided his army into two corps, the one retreating along the Weser to Cassel, and the other to Munster: in this case, it would have been impossible for the allies to have made any movements to fall upon the left wing of the French; nor could the combined army have advanced a step, without exposing itself its right flank to the division of the French, retreating towards Munster.

I have confined myself, in my observations on the seven years war, to arguments, I believe, never before used, and which demonstrate how much the principles I have explained in the first part of this work may assist us in our judgments on military events; for, it was not my intention to write the history of a war, which has already employed celebrated pens. This memorable period, when so many bloody battles were fought, having furnished numerous opportunities of making experiments, which have unfolded and brought to perfection the system of fire, has laid the foundation of an order of things, in which battles will entirely disappear from the theatre of war. It is a fact, that since the seven years war, we have had no battles to be compared to those so often fought during those years.

CHAP. III.

Observations on the Wars which have occurred since the seven Years War, and on the first Events of the War of the French Revolution.

THE principle of the base is confirmed by the wars of the Russians and Turks.   It is impossible for the Russians to maintain themselves beyond the Danube; and it is still doubtful whether, in spite of the ignorance and bad military constitution of the Turks, the conquest of their empire in Europe be not chimerical.   It seems to me, that the want of provisions will always prove an insuperable obstacle to the marching of an army to Constantinople over the Hæmus.

The war for the Bavarian succession does not deserve the denomination of war.   Negotiations never ceased during the whole continuance of it.   Joseph II. was inclined to it, yet without a determined will to incur the dangers of it. The zealot Maria Theresa, whose conscience was, perhaps, tormented on account of the

blood shed in the seven years war, was decidedly against it; and the philosopher Frederic considered war as a wicked folly, when it was not absolutely necessary. Thus the parties had no
mind to fight, and only opened the campaign
to prove that they were ready.

The project mentioned by Frederic, in his
history, of carrying the war at this period by the
back of Moravia to Presburg, will bear no examination; for, the operation having no base,
would not have succeeded, even after the taking
of Olmutz.

Prince Henry's march through the difficult
defiles of Gabel, to penetrate into Bohemia, is
an effect for which we see no sufficient cause.
There are easier roads by which that kingdom may be entered There was, therefore,
no motive for this operation, and in general
there is nothing in this war that deserves particular attention.

The American war, of little consequence in
some respects, is peculiarly remarkable and
important, as a new political and military æra
I shall consider it here only in the latter point of
view: there were no great battles; nothing but
slight actions; it was a war of light troops, as
we shall see.

I do not mean to dwell on the inconceivable

faults committed by the English generals. How often was it in their power to put an end to this war, by a decisive attack on the handful of men who defended the liberty of America! They were, no doubt, prevented by a destiny, more powerful than men, and which attains its ends in spite of their designs. The manœuvres of the American general at Trenton and Prince-town are masterpieces. They may be deemed models for the conduct of a general, supporting a defensive war against a very superior enemy. General Washington fell, at different times, on the rear and flanks of the enemy, threatened his supplies, and constantly reaped the fruit of operations at present truly efficacious.

Montgomery's enterprise, and Arnold's march into Canada, merit praise. The plan for making Lord Cornwallis prisoner, at Yorktown, was well laid, but must have occurred to every one's mind. What, however, renders this war remarkable in a military point of view is, that the first use of *Tirailleurs* may be dated from that period, and that the American soldiery were the first troops employed in that manner.

The British cabinet laid down a plan that deserves attention. It was to form a junction between the army at New York and that in Canada, and by that means to separate the

s 4

American colonies from one another Its not succeeding was entirely the fault of Sir William Howe, who marched to Philadelphia instead of directing his operations northward towards Hudson's river. General Lloyd states an infallible means of subduing New England, "with " the conquest of which," says he, " all North " America will again fall into the hands of the " English, for the republican spirit of insurrec- " tion exists only in New England." He says that the English should occupy Boston and Rhode Island, and detach corps from those points into the interior of the country, while another army should march from Canada along the Hudson In this case it is certain that all the operations of the Americans would be eccentric If they advanced to meet one of those columns, they would be taken in the rear by the others. However, it should seem for the complete execution of this plan, that the English should also be masters of New York.

It appears to me that the English in this war should have occupied at least all the principal ports. This measure alone would have subjugated the American people, who are all traders. Cut off from every communication with Europe, those colonists would have seen the impossibility of supporting themselves. Let me not be told that the English blockaded their ports; such an extent of coast is not to be blockaded. But had garrisons been established in the ports, the American navigation would have been stopped at its source. Twenty thousand men would have been enough to effect this, and the English kept up at least thirty thousand in America. Ports too unimportant to be occupied they might have destroyed, and by stationing cruizers along the coast, they would have completely put an end to the communication with France, which alone supported the energy and hopes of the Americans during this war.

A colony accustomed to the luxury and to the supplies of the mother country may be soon subdued by cutting off its trade with the latter. The Americans, who could not provide themselves with what their necessities required, would have fallen into extreme want. A general cry would have arisen against the authors of the rebellion; and the nation would have begged for peace on their knees. To gain such advantages, all the English had to do was to take possession of Boston, Rhode Island, New York, Philadelphia, Norfolk in Virginia, and Charlestown and Savannah in the South: nor are we to imagine that it would have been very difficult to master those places; for in this war the English and Hessians always took whatever place they chose to take. And should the possession of the ports have been insufficient, the English, according to my calculation, would still have had a reserve of ten thousand men, with whom, by occupying the mountains from Canada, they would have reduced the Americans to their narrow valleys, and completely subdued them. I could easily give a minute view of these operations, were it not enough to observe that we are here in no respect speaking of a warlike people, such as the Europeans; that, in regard to the Americans, it would have been possible without danger to make a thousand exceptions to the great principle of the base displayed in this work, whereas the slightest deviation from that principle with a nation really military may be attended with the most serious consequences. All things considered, the Americans are indebted to the inadvertencies of the English generals for the first steps towards their liberty; and for the confirmation of it, to the arms of France, whose interest it was to support them. What they owed to their own talents and bravery amounts to little or nothing.

Before I conclude this chapter, I cannot but

take a glance at the beginning of the war of the French revolution, in which the Prussians took a part. It is, however, by no means my intention to consider that war in a political point of view. I confine myself here to military inquiries, and in a military view alone I mean to examine a very small portion of a memorable period, the history of which it is not the part of contemporaries to write.

In this war we perceive new developements of the modern system. There is an uninterrupted progress from the seven years war to this, however different those eras appear in other respects. I shall go immediately to the moment when the Prussian army arrived on the frontiers of France. This campaign is assuredly no proof against the military talents of the commander in chief of the Prussians; because it was too evidently conducted against all the rules of war. Can it be supposed that the Duke of Brunswick was ignorant of what every body knows, that hostile armies and fortresses are not to be left in the rear; that a multitude of men cannot be transported through the air from one place to another, and that those men must be fed? Even if the Duke had not given, in a thousand instances, proofs of his military talents, the errors of this campaign could never be imputed to him, from the very circumstance of their being too great and too evident. We must therefore have recourse to other causes in accounting for so singular a conduct; and those causes are very naturally found in the political intrigues from which a powerful effect was expected.

It appears that the Duke had several times in the council stated the necessity of establishing a base by the conquest of several fortresses, in order to advance in a regular manner; which was as much as to say that if this precaution were neglected, he did not answer for the consequences. The Duke, to all appearances was constrained to follow opinions different from his own; and as the march of the Prussians into Champagne was undertaken on erroneous political calculations, no judgment can be formed of it in a military point of view It is a monster in the military art, unlike every thing of its kind.

It was not the same case with General Dumourier: who, commanding the French army could have no motives of a political nature to prevent his acting according to the rules of war If he had, they are nothing to the point, for it was his duty not to suffer his political opinions to have any influence over his military operations, which were to serve for the defence of his country. We are therefore warranted to consider his conduct entirely in a military point of view, and to inquire impartially whether his campaign in Champagne, and which he gives as a model of defensive war, really merits such praise. Not that I absolutely blame a general for speaking well of himself; there is a certain frankness which becomes a warrior in praising, even in his own achievements, what is worthy of praise.

At the time when the Prussians entered the territory of France, the French armies were extremely well posted. Kellerman was encamped near Metz, Dumourier near Sedan; there was another army near Valenciennes, and Custine was in the neighbourhood of Landau. These are precisely the positions pointed out by General Lloyd for defending the frontiers of France in those quarters. Kellerman, particularly, was excellently well situated at Metz, having it in his power to crush the Prussians if they advanced, as in fact they did, between Thionville and Sedan: and yet Dumourier withdrew Kellerman from Metz, to form a junction with himself by a round-about way: the first capital fault.

What effectual services might not Kellerman

have done had he acted from the position in which he was? He might, by the superiority of his forces, have attacked and overthrown the small corps of Austrians who were bombarding Thionville; and, then, leaving one detachment to check that corps, and another to oppose any enterprises of the garrison of Luxemburg, he might have marched the rest of his army to Treves and Coblentz, and taken those two places. This would not have been difficult, for there was no garrison in the latter of those towns.

The Austrian corps, indeed, driven from Thionville, would not have retired to Luxemburg, which was of itself very strong, but, no doubt, to Treves and Coblentz. Still, Kellerman would have been able to stop the convoys from Luxemburg to the Prussian army. What would have been the situation of this army, which must have been in danger of being cut off from Coblentz, while Custine would have been making himself master of Mayence and Frankfort? If, in addition to this, Dumourier had, as he might have done, used the means offered to him of then conquering the Low Countries, not only would the Prussians have had no retreat, and found themselves in a most critical position; but they would have been

stopped in their march towards Paris, and reduced to a dangerous defensive war, which,
too, would probably, have been unsuccessful.
We may judge, then, of the importance of diversions made on the rear of the enemy, and
of the efficacy of this kind of operation.

Dumourier, on taking the command of La
Fayette's army, called a council of war. General Dillon proposed a diversion in the Low
Countries, which, considering the weakness of
the Austrians at that juncture, would have succeeded: and had Dumourier employed his army
in that enterprise, and Kellerman at the same
time driven the Austrians towards Luxemburg, and
thence as far as Coblentz, the retreat of the Prussian army across Ardennes would have become
impossible. What a campaign would this have
been for the French! And how poor does what
they did then appear in comparison!

This campaign would have been equally successful, even if the Prussians, on being informed of Kellerman's march, had retired
towards Luxemburg in time to prevent their
being cut off; for, necessarily too much occupied by Kellerman's army, and alarmed at
Custine's diversion on Frankfort, it would not
have been in their power to interrupt Dumourier's operations in Belgium. Nor was Clair-

fait, who had scarcely ten thousand men, and who besides might have been cut off on the side of Givet, dangerous to the French general. Dumourier having a perfect base on the line of fortresses extending along the Meuse, might have marched unimpeded to the Rhine. The operations of Kellerman and Custine, well concerted, would have compelled the Prussians to march in haste into Westphalia, that they might not be in danger of having their communication with their own country intercepted, and the French would have been, from that period, masters of the course of the Rhine from Bâle to Holland, to obtain which afterwards cost them seven campaigns, and torrents of blood.

Instead of this, what did Dumourier do? What all generals do, who are more guided by sensible impressions than by reflexion. He opposed the enemy in a direct manner, and an enemy against whom the unorganized troops he then commanded were incapable of making head. His only excuse is, that he was afraid of incurring the censure of the Parisians, if he had withdrawn from covering their town, to make a diversion into Belgium. All that can be said to this is, that it is very unfortunate that a leader of armies should be influ-

l

enced by *badauds**. Still, Dumourier might have
recovered his advantages in another manner;
and, as the Prussian army had advanced into
the interior of France in such a way, that circumstances compelled a retreat, the French
general ought not to have allowed that retreat
to be effected so quietly. In general, there is
nothing to be praised in this campaign, on
either side, but Custine's diversion to Mayence
and Frankfort: all the rest is stamped with absurdity and want of foresight. However, Custine displayed no talents beyond the occupation of Mayence; for, his operations on the
right bank of the Rhine is even beneath criticism. We may remark, too, that the taking
of Savoy, in this campaign, which might
seem a brilliant event, was entirely owing to
the cowardice of the Piedmontese troops, who
deserted their country.

Dumourier set out brilliantly in Belgium, by
the action at Jemmappes. I say *action*, and
not battle, as the whole loss on both sides,
killed, wounded, and taken prisoners, scarcely
amounts to five or six thousand men: nor was

* A nickname given to the Parisians, expressive of their
curiosity and eagerness for novelty. The original sense of
the word is simpleton.—COMMENTATOR.

T

the number of the combatants very considerable;
the Duke of Saxe-Teschen had not twenty thousand men, Dumourier had about sixty thousand.
Indeed, this superiority was fully sufficient to have
enabled him, on the one hand, by manœuvring
skilfully, to have forced the Austrians to evacuate Belgium without a battle; and, on the
other hand, to give him no reason to boast of
a victory, for which he was indebted solely to
numbers. It is in a contrary case that success
gains glory for a general. How poor does the
action of Jemmappes appear, compared to the
battles of Frederic in the seven years war!
But all these considerations were nothing to
Dumourier who wanted to gain a battle.
However, during the action, he displayed courage, presence of mind, and military talents.
The bravery of his troops deserves no less to
be praised. From this moment, the Germans
began to abate the contempt which they
thought they had a right to entertain of the
French troops, since the seven years war.
They were afterwards furnished with still better occasions for curing themselves of this pre-
judice, and, according to all appearances, it is
now suppressed for ever.

The military career of Dumourier terminated in the commencement of the campaign of 1793. In the preceding winter, he had committed a capital error in not exerting all his strength and talents to drive Clairfait beyond the Rhine. He exposed his flank by Belgium, while he undertook the conquest of Holland. This error in calculation, which, in the modern system of war, can never fail to fall with violence on him who commits it, was, in fact, the source of Dumourier's disasters: yet, never did he display more ability than in the celerity of his exertions to repair them. The attack, however, of his adversary being also very active, and well planned, the resources of genius displayed by Dumourier in his defence availed him nothing.

I have dwelt on the conduct of a general, who, after having reanimated the relaxing spirit of his countrymen, and laid the foundation of all their future successes, is become a stranger to his country. From this circumstance, and that of his having publicly entered the lists, pen in hand, to challenge criticism, I thought myself warranted to speak of him with freedom. He, no doubt, possesses very various and extensive knowledge, but it is not digested into a system: he has imagination, fire, talents, but no character; in a word, never was there a Frenchman more a Frenchman than Dumourier. With him I conclude my reflections on the war of the French revolution; for I do not think the time yet come for speaking of other operations.

CHAP. IV.

What are the natural Bounds of the great European States?   What Inference, in respect to future Wars, may we draw from the Determination of those Bounds?   A Military Problem proposed.   Conclusion.

IN all probability wars will yet be waged for a long time: they will be waged till repeated experience has shown that States have nothing more to gain from one another; till there is nothing more to conquer or to share; in a word, till the great nations now existing have attained their natural limits.   It were to be wished, for the repose of the world, that all possible pretensions on this head were settled at once in an amicable manner.

It will be asked, what then are the limits which nature appears to have prescribed to the great empires now in being?   We have seen that, in a military point of view, only the sea, mountains, and rivers, can be called natural limits; for, they interrupt the communications, and increase the difficulties of them. Nations and languages are here of no consideration. In our times, we see many instances of several people speaking different languages subject, without inconvenience, to the same government. One of the reasons of this is, that each State has formed a military bound, within which all is subjected to it, because it can there collect more materials of war than the neighbouring Powers. It is the nature of the country, much more than the inhabitants of it, which now constitutes political bodies: a striking emblem of the material mode of thinking in this age.

According to these principles, I find, that the Rhine can, by no means, form a natural boundary between France and Germany. The left bank of the Rhine, as far as the Meuse, is certainly within the military boundary of the Germans, and, in that space, their action must repel the French re-action. I put the present political situation of those two Powers out of the question, and, for a moment, consider Germany as only forming a whole. It is evident, that this formidable body would, by its masses, crush those which France might bring against it, even considerably within Belgium. The reason is to be found in the nature of the country; for the sea is too near it, and the land extending from the Meuse to the sea too narrow, to allow of collecting there so great a quantity of the materials of war, as the Germans would draw from their country, which they would have entirely behind them.   Thus, the German lines of operation would be straight, and consequently shorter than those of the French. The French armies, in order to face the Meuse, would have to make a wheel to the right strategically.   This movement would prevent their drawing directly from behind them the means of carrying on war.   They could not supply them, but from the interior of France, whose old frontiers form an angle with the position, which, in that case, its armies would have upon the Meuse.   This would, of course, lengthen the lines of operation; for, the nature of the country would prevent the French receiving their convoys by their flanks across the Ardennes.   But we have seen, that the pression of military forces diminishes in proportion to the length of the line of operation: the pression of the German forces would, therefore, be more efficient, particularly as having behind them a country as extensive as rich in materials for war

But, if the Germans pass the Meuse, they

T 4

lose their advantage. As soon as their armies should be forced to wheel to the left, in order to march perpendicularly to the French frontiers, they would find themselves exposed to the same inconveniences which the French might have experienced previously in respect to them. It appears, then, that while France and Germany make war upon each other, Belgium will be the theatre of it. If Holland be in alliance with the Germans, the French would find it so much the more difficult to preserve the whole of Belgium: if, on the contrary, it be with France that Holland is allied, the Germans would only have a greater number of points by which they might injure the interests of France. Holland is open on the side of Germany; its defence would cost the French dear, and would weaken them considerably in Belgium. Had it not been for the peace, concluded in the year 1795, I am persuaded that the Prussians would have reconquered Holland in less time than a campaign. I am aware that France may erect a multitude of fortresses on the Rhine; but, as I am convinced that the superiority of warlike masses is ultimately what decides success in war, these fortresses would but render the passage of the Rhine more difficult, without preventing it. Besides, would

France have time to raise those enormous structures before a new war should break out? But all this reasoning depends upon the supposition that Austria and Prussia would enter into a new coalition, which is become almost impossible; and the actual state of things gives France a great advantage over the German Powers.

France, on the side of Spain and of Italy, has insurmountable bounds, the Alps and Pyrenees. It is to be presumed, that the Spanish peninsula will, sooner or later, form but a single State: it will, in the end, be the same case with Italy; and then the Austrians will never be able to defend their new acquisitions in the territory of Venice against the State united*. The frontiers of this detached quarter will be surrounded on all sides by those of the Italian Empire; the operations from the latter will, therefore, be concentric, while those from the former will, on the contrary, be eccentric. Thus, Austria will, in the course of time, be dispossessed of this territory; but, on the other hand, the frontiers of the Tyrol will always be an insuperable barrier against the attempts of Italy.

---

* It must be observed that this work was written previous to the late occurrences in the Austrian dominions.

COMMENTATOR.

To compensate for this loss, it is impossible but that Austria must, one day or other, acquire the possession of the whole course of the Danube to its mouth; because the nature of things will, in this, second the military plans of the cabinet of Vienna. The Austrians take in flank all the operations of the Russians against Turkey: so that Russia can look for no success on that side, without the concurrence of Austria. The Grecian peninsula appears, besides, to be safe behind the mountains which divide Thrace and Macedonia from Bulgaria and Servia; and, Turks as the Turks are, the enterprises of the Austrians and Russians against this part of the Ottoman Empire would still be fruitless. When once Austria shall be in possession of all the vallies of the Danube, she will, by separating Russia from Turkey, put an end to the wars between those two Powers, and the independence of the Turkish peninsula in Europe will be secured; for, it will be the interest respectively of Russia and Austria, to prevent each the other to conquer it.

The country of the Danube, though inhabited by Germans, seems not to belong to Germany. The reason is, that that river falls into the Black Sea; whereas, the other rivers

3

of Germany disembogue into the North Sea and Baltic.

The military boundary of Russia will always hold to the north of the Carpathian mountains; and, in the event of a rupture between the two Powers, Austria would find it difficult to preserve its Polish possessions beyond those mountains. But, in the present state of things, the best plan she could adopt, in case of a rupture with Prussia, would be to act in two columns on the right and left of the Vistula, (which will be easy while Warsaw remains unfortified,) and further, to abandon the north of Bohemia to the inroads of Prussia, as it would be impossible for her to defend that country.

The Prussians would find it no less difficult to protect Eastern Prussia against the attacks of the Russians; for, the frontiers of the Russian empire enveloping that country, all its offensive operations on it would be concentric, and those of the Prussians in defending it, eccentric. It would be necessary for Prussia to raise a chain of fortresses from Memel to Warsaw, including particularly the latter, which ought to be made the bulwark of that monarchy, on the east, as Magdeburg is on the west. It appears that the Russians may likewise extend advantageously their military boundary, even on the west of the Vistula, as far as the Oder. All the country, watered by the latter river, would always belong to the Prussian empire, even if it lost that on the Vistula.

But, in all cases, and without a possibility of its being prevented, Prussia will find indemnities in the north of Germany. Sooner or later, this extensive country will be divided into north and south, the one under the dominion of Prussia, the other under that of Austria. Prussia, once mistress of the northern part, could never be expelled from it by any Power: she would easily extend her influence, and even her dominion over Holland, and the continental territory of Denmark, especially if the latter monarchy, continuing in the leading-strings of England, is incapable of rising from her political nullity, by giving herself value to the productions intrusted to her by nature, the only means by which a State can lay the foundation of real power.

We may conjecture, then, on probable grounds, that the time will come, when Europe will consist of the following States: Spain; France; Italy; Switzerland, the mountains of which will preserve its independence; the Austrian empire, including the south of Germany and the course of the Danube; the empire of Prussia, including with its present possessions the north of Germany; Denmark; Sweden; Russia; the British islands; and European Turkey   To compose these twelve States, I reason on the principle, that the great ones now existing will grow still greater, and the small ones be swallowed up   Whatever political constitutions those States shall adopt, they must be lasting, from the physical and moral nature of things.   This new division of Europe will be favourable to perpetual peace; for geography teaches us that each empire will then be bounded by the limits to which it may pretend, and beyond which it will not be its interest to go.

Before I conclude, I beg leave to propose a military problem.   Will the Tartar nations ever be able again to overrun and subjugate Europe? I cannot undertake to answer this question: there are not, in my opinion, data enough to conclude either for or against the affirmative.   The Tartars, issuing forth in a multitude, advance with incredible velocity, almost without provisions: but would they be able to take our fortresses? I do not know: they would at least ravage the country in such a manner, that they would compel it by famine to submit. Tamerlane subdued towns. We are not yet sufficiently acquainted with the military art of those nations. I wish this question were examined.

A certain number of my principles will, no doubt, be granted, but others of them will be treated as paradoxes. I shall cause surprise in maintaining that the parallel defensive war is to be abandoned; that a battle may always be avoided; that a defensive war ought to consist of offensive diversions; that light infantry is sufficient in the modern system of war, &c. I submit these principles to the judgment of the Public, while I declare myself convinced of their truth. My style, I trust, will not be too strictly criticised: I have only aimed at being clear; in which, from the arrangement I have made use of, and the simplicity I have endeavoured to preserve, I flatter myself I have succeeded.

THE END.

Printed by C. Mercier and Co.
Northumberland-court, Strand, London.

Pl.1.

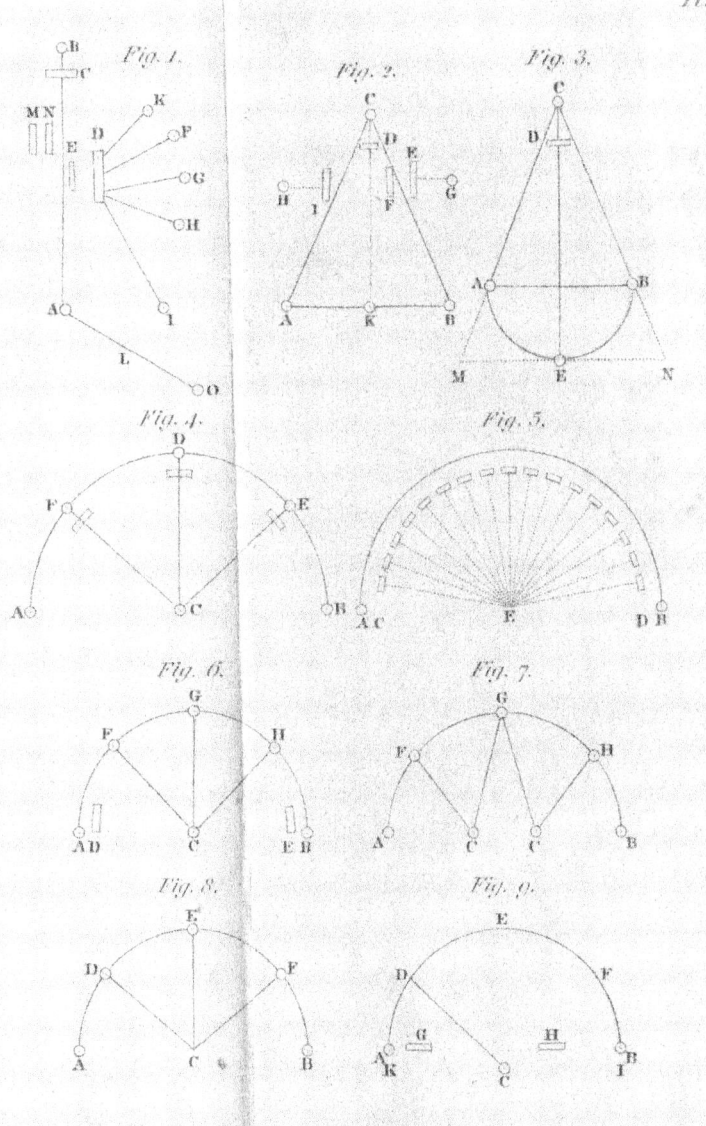

Fig. 1.   Fig. 2.   Fig. 3.

Fig. 4.   Fig. 5.

Fig. 6.   Fig. 7.

Fig. 8.   Fig. 9.

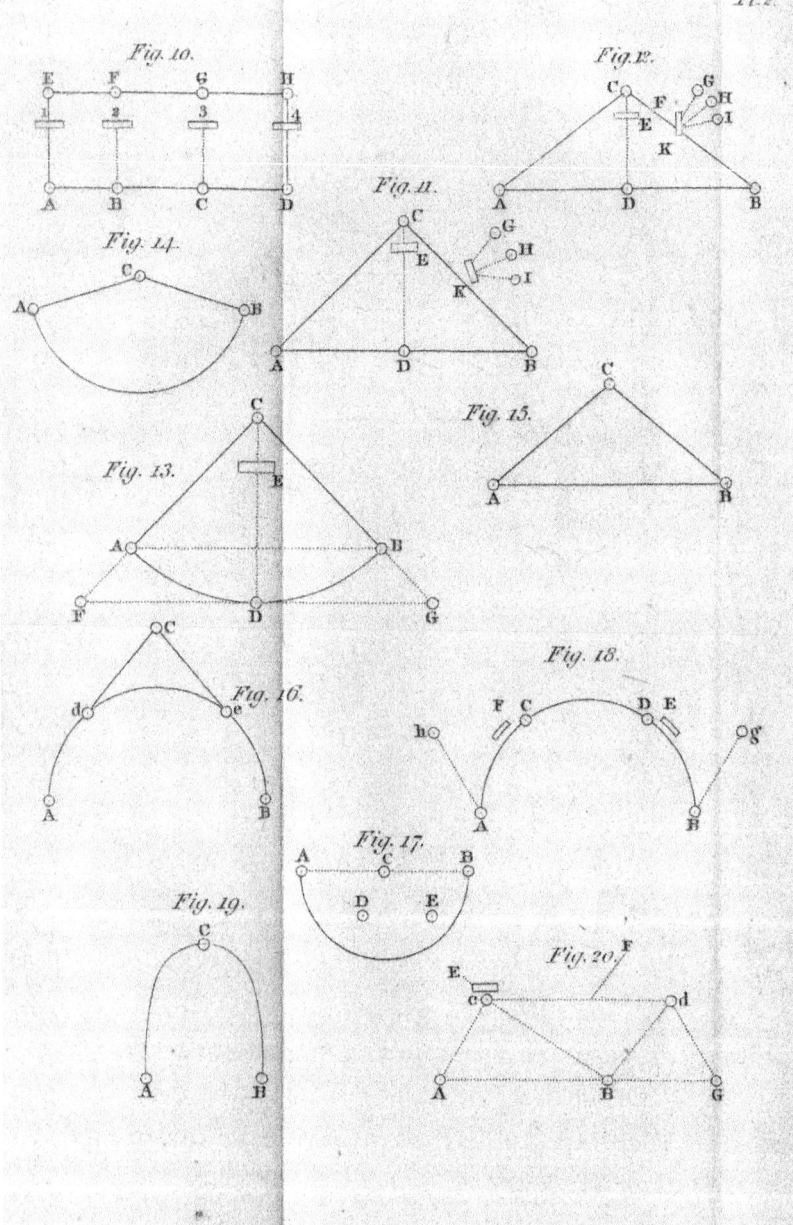

Pl. 2.

Fig. 10.

Fig. 12.

Fig. 11.

Fig. 14.

Fig. 15.

Fig. 13.

Fig. 16.

Fig. 18.

Fig. 17.

Fig. 19.

Fig. 20.

Pl.3.

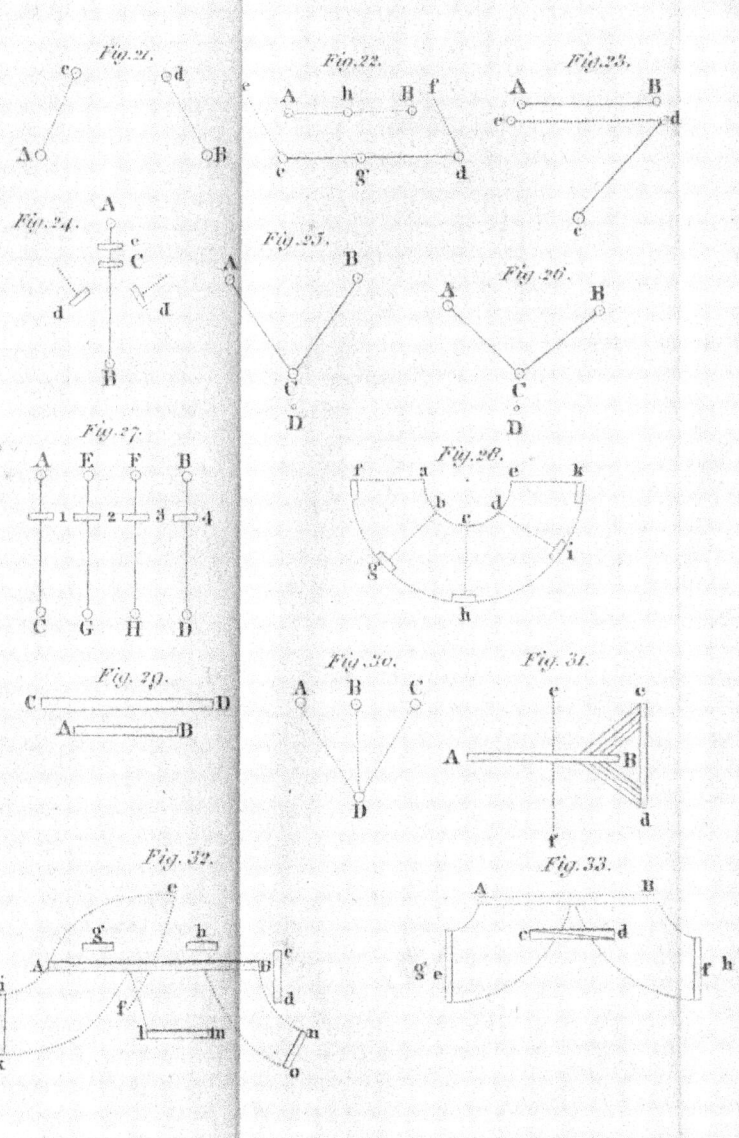

Fig. 21.

Fig. 22.

Fig. 23.

Fig. 24.

Fig. 25.

Fig. 26.

Fig. 27.

Fig. 28.

Fig. 29.

Fig. 30.

Fig. 31.

Fig. 32.

Fig. 33.

Pl. 4.

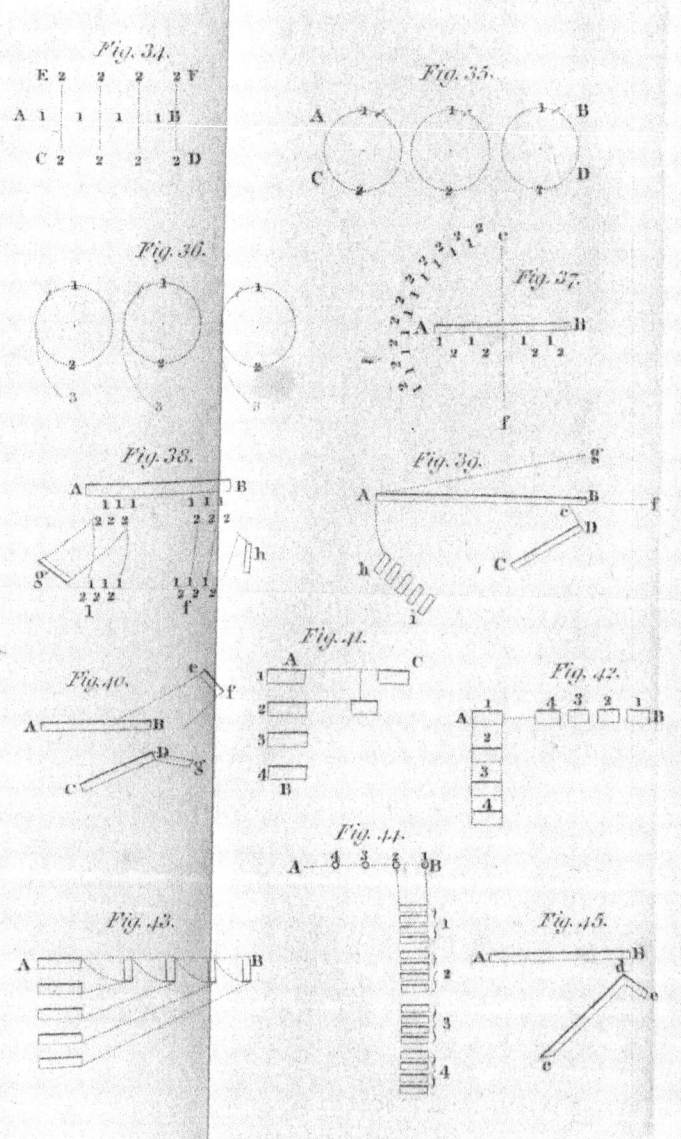

Fig. 34.

Fig. 35.

Fig. 36.

Fig. 37.

Fig. 38.

Fig. 39.

Fig. 40.

Fig. 41.

Fig. 42.

Fig. 43.

Fig. 44.

Fig. 45.

Pl.5.

Fig. 46.

Fig. 47.

Fig. 48.

Fig. 49.

Fig. 50.

Fig. 51.

Fig. 52.

Fig. 53.

Fig. 54.

Fig. 55.

Fig. 56.

Fig. 57.

Fig. 58.

Fig. 59.

Printed by Printforce, United Kingdom